CHERNOBYL: A DOCUMENTARY STORY

Chernobyl: A Documentary Story

Iurii Shcherbak

Translated from the Ukrainian by Ian Press

Foreword by David R. Marples

Canadian Institute of Ukrainian Studies
University of Alberta
Edmonton 1989

Canadian Cataloguing in Publication Data

Schcherbak, Iurii. 1934–
Chernobyl: a documentary story
Translation of: Chornobyl': dokumental'na
povist'
ISBN 0-920862-64-0 (bound). —ISBN
0-920862-65-9 (pbk.)
1. Chernobyl nuclear accident, Chernobyl',
Ukraine, 1986. I. Canadian Institute of
Ukrainian Studies. II. Title.
TK1362.S6585313 1989 363.1'79 C88-091524-2

Typesetting by Bruce Hartshorne Consulting Limited

Distributed by
University of Toronto Press
5201 Dufferin Street
Downsview, Ontario
M3H 5T8

Contents

Foreword

David R. Marples

Yurii Shcherbak's *Chernobyl* was published in the Soviet monthly journal, *Iunost*, in two issues in the summer of 1987, and also in the Ukrainian journal, *Vitchyzna*, in the spring of 1988. This English version appears as a result of an agreement negotiated between the Canadian Institute of Ukrainian Studies (CIUS) and the USSR Copyright Agency in Moscow. Although a second volume is now being issued by Dr. Shcherbak, it was felt by CIUS that the first is complete in itself and that its appearance before the Western public should not be delayed further. This remarkable little book represents our first eyewitness testimony to the events of and succeeding the Chernobyl disaster of April and May 1986.

To the Western mind, the word Chernobyl brings to mind specifically the week or several weeks that followed the nuclear accident of April 26, 1986. Our attention span is guided by the media, often by the headlines in several selected newspapers. Chernobyl as an event was notable in dominating those headlines for perhaps longer than any other event in the memory of a postwar generation. Yet, in early May, its extensive coverage ended. It had other effects: anti-nuclear groups in the West were revitalized by the disaster. Several countries whose governments had wavered over the choice of nuclear or non-nuclear energy finally gravitated toward the latter course. Thereafter, the main concern in the West about Chernobyl was the state of imported food in North America, and in Western Europe, the radioactive fallout and its effect on this same food chain.

For the Soviet Union, however, the real drama of Chernobyl had yet to unfold. A dramatic battle had begun to counter the fallout of the raging atom that continues today. It is being fought by various means, some successful, others less so. While some areas were successfully decontaminated, others that appeared to be well beyond the danger area were found to be irradiated. The population appeared to be often confused, fearful of the effects of radiation, known and unknown. It is symptomatic that as late as the summer of 1988, the two main newspapers in Kiev began to publish figures on the radiation background, to assure citizens that it was within the norms of safety.

Chernobyl occurred during the General Secretaryship of Mikhail Gorbachev, a leader and statesman who had quickly earned the respect of the West, particularly through one of his twin policies, glasnost, or openness (the other is perestroika, or restructuring). Glasnost, somewhat belatedly, played its own individual role in the sphere of nuclear energy, particularly after the summer of 1987, when a concern for the ecology spilled over into the nuclear sphere. In Ukraine, through the vehicle of the weekly newspaper *Literaturna Ukraina*, the Ukrainian Writers' Union questioned the viability of a programme for nuclear power development that had not taken into account its effect upon the natural environment. It was felt that the republic lacked the necessary water supply for a major capacity expansion, while in areas such as Chyhyryn, the planned nuclear power plant was to have been built in a famous historical area, the former seat of the seventeenth-century Hetman state.

By the fall of 1988, the Soviet nuclear power programme had been modified considerably. At the time of writing, it has just been announced on Radio Moscow that a Commission headed by Vice-President of the Academy of Sciences,

Evgenii Velikhov, has concluded that the Crimean nuclear plant should not be built because it is located in an area of high seismic activity. Velikhov's name appears in the pages of this book. Public feeling in the republic is so sensitive about the subject of nuclear power plants that several other plants in Ukraine have also been shelved or abandoned: Odessa, Kiev, Chyhyryn and Kharkiv.

Another prominent name that features regularly in these pages is that of Valerii Legasov, the First Deputy Chairman of the Kurchatov Institute of Atomic Energy and the head of the Soviet delegation to the International Atomic Energy Agency in Vienna in August 1986. In April 1988, Dr. Legasov committed suicide. His memoirs, published posthumously in *Pravda*, reveal a deep concern for the way in which nuclear power plants were being built and operated. His tragic and untimely death cannot be divorced from the accident at Chernobyl, as Shcherbak reveals in his interviews with the late scientist in these pages,

This is not to say that Yurii Shcherbak and his colleagues are opposed to nuclear power or are fundamentally at odds with the Soviet energy programme per se. As the text reveals, Chernobyl has had a profound psychological impact on the Soviet people. Many feel, like Shcherbak, that if pitfalls are to be avoided in the future, its lessons must be learned. Through the interviews held with participants in the Chernobyl tragedy throughout this book, Shcherbak reveals both the good and the bad. The underlying statement, which surfaces in the final chapter, is that this is "The Last Warning." Another mistake with the atom could be the last. It is fair to say that generally this attitude pervades the Soviet rather than the Western public, but that is not to deny its validity.

Shcherbak himself is a trained doctor, and the author of some nine acclaimed novels. He is also chairman of a group within the Ukrainian Writers' Union known as *Zelenyi svit* (Green World) that is concerned with the protection of nature. This combination of talents and interests has enabled him to produce a unique historical document; an eyewitness testimony to Chernobyl that is interspersed with the author's frank and perceptive comments. The immediacy of the disaster is illuminated in these pages by those who lived through it: firemen, doctors, scientists, party and government workers, helicopter pilots and journalists. There is the contrast between those collapsing from radiation sickness after fighting the graphite fire in the building of reactor No. 4 and the final chapter in which the author comments on the simple joys of lying on an uncontaminated bank aside a peaceful river. We can either take care of this land, Shcherbak seems to say, or see it turned into an unpopulated and uninhabitable desert.

Translator's Note

Iurii Shcherbak's *Chernobyl* is an important event in Ukrainian, Soviet and world literature. The views and reactions of people of all walks of life are represented here, as they spoke, shaped by the art of a considerable writer. I have done my best to give a reasonable rendering of the text as printed in two issues of *Vitchyzna*, including the preliminary section *Vid avtora*, which follows this as the Author's Preface and explains better than I could aspects of the history of the work and the writing of Book 2.

Some preliminary notes on the translation are appropriate. First, the title. It is, we all know, Chornobyl, but the world knows it as Chernobyl and I have decided to use this form. Prypiat, however, remains Prypiat and all other place and personal names follow CIUS style. Secondly, abbreviations are either expanded or left as they are, with explanation as appropriate. Where an acronym has an internationally recognized form, e.g., CPSU for Communist Party of the Soviet Union, I have used that. I have used USSR for the Soviet Union and UKrSSR for the Ukrainian Soviet Socialist Republic. Thirdly, to make matters absolutely clear, this translation has been made directly from the Ukrainian text, excepting the Russian song by Vysotsky and a reference to the Russian texts of Gubarev's *The Sarcophagus* and Gogol's *Mirgorod*. I am very grateful to David Marples and CIUS for resolving the difficulties with which I left them.

The experience of Chernobyl is unique in the history of the world in revealing to us the potential scale of a catastrophe involving the peaceful use of nuclear energy and the dangers of the appeal of 'technology for technology's sake.' There is much to learn, and Shcherbak, in his insistence on the need, everywhere, for openness and international co-operation, sets us well on the way to a more secure future.

Ian Press
Queen Mary College
University of London

Author's Preface

There is one particular question among the many with which readers of the documentary story Chernobyl unfailingly confront me, whether it be in letters or in the course of numerous encounters: Why did I first publish this story not in Ukraine but in Moscow, in the magazine Iunost?

And almost always the questions simultaneously carry within them a sort of answer: 'We, of course, understand that in Ukraine you could hardly have managed to publish it in full, but... '

I consider it a duty to give a truthful answer, bearing in mind that the story reaches the Ukrainian reader after a year's delay. Like any Ukrainian writer, I bear a sense of deep responsibility for my native language and literature and their development, and so it is not appropriate to avoid fundamental, burning and highly topical questions. Particularly today, when we are living through a moment of truth.

First. The magazine *Iunost* played a significant role in my literary life: it was precisely in its pages, starting in 1961, that my first articles were published, and in 1966 the editor-in-chief, Boris Mikhailovich Polevoi, supported my candidature for the Union of Soviet Writers. I shall always remember this with pride and gratitude.

And so, when at the VIII Congress of Soviet Writers in

Moscow in June 1986 the editor-in-chief of *Iunost*, Andrei Dementiev and his editorial colleague Iurii Zerchaninov came to me and proposed that I write a documentary story about Chernobyl for the magazine, how was I to refuse? And was it necessary for me to refuse? Particularly since not a single Ukrainian magazine had at that time rushed to me with such a proposal. Moreover, I am not at all convinced that I would have written this story had it not been for the initiative of *Iunost*. I had given no thought to it over that anxious summer, and could quite easily have limited myself to articles in *Literaturna Ukraina* and *Literaturnaia gazeta*.

Secondly. Well aware that it was a very important all-Union vehicle (the circulation of *Iunost* amounts to over three million copies!), I very consciously wrote the story precisely for this magazine, wanting its numerous readers in the remotest corners of our land and abroad (the magazine is read all over the world) to find out as quickly as possible about the true course of events and the real dimensions of the national calamity which befell us in April 1986. I became convinced that I had done the right thing when, during my recent journey to Venezuela and Peru, I met people who had read the story; and this is also clear from the letters which I receive from all over the Soviet Union and other countries.

And, thirdly and last. I will not go against what I really feel. The further my work on the text proceeded, the more clearly I understood that its fate in Ukraine might not be an easy one. I have considerable experience of work in Ukrainian literature and, in the past, have experienced for myself that braking mechanism which in the years of stagnation I would automatically apply. I could imagine how. I knew precisely in which 'hot spots' of the text someone's cold and pitiless red pencil might pass, smoothing over, embellishing and softening that which could not be softened if one wanted to bring back a sense of worth to oneself and one's words.

At that time, in the summer of 1986, the struggle for glasnost was only just beginning to unfold, and many areas, Chernobyl among them, remained beyond the bounds of criticism. The tradition of regulating and rationing the

Truth, as if it were something in short supply, was still alive. One thing was permitted 'in the centre' and another—far less of it and far worse—in the localities. And I was convinced: there was one Truth for everyone, our Truth, the Soviet Truth. There wasn't and couldn't be a 'republican,' 'regional' or 'district' truth. This was an absurd fiction of people who were against a breath of fresh air, who naively felt that it was possible, at the end of the end of the twentieth century, to get rid of awkward problems which had developed in our society by means of prohibitions and silence... Perhaps I was wrong in my pessimistic forecasts regarding the amount of glasnost in Ukraine. Perhaps my story would have been published without the slightest cuts. Who knows...

Upon reflection, when I gave my work to the magazine *Iunost* I had before me the example of Oles Honchar, who first published his story of warning *The Black Ravine* in *Moskovskaia pravda*. In their time the same thing was done by Chingiz Aitmatov, Vasil Bykov and Ion Drutse.

And all the same, even as I tried to persuade the reader of the rightness of my choice and decision, all the time I felt pangs of conscience: Chernobyl had to become a fact of Ukrainian literature.

And I am very grateful to the magazine *Vitchyzna* for proposing to publish the Ukrainian text of the story, all the more as a part of the sections was originally written in Ukrainian.

At this moment I am working on the second part of Chernobyl, and I hope that this year it will be published simultaneously in the pages of *Vitchyzna* and *Iunost*. A great deal of material has been gathered: the testimony of witnesses, letters, recollections and documents. I want to believe that this will give me a chance to reveal unknown aspects of the accident to the reader, to recreate objectively the atmosphere of those tragic days, to demonstrate the greatness of the people who in improbably difficult conditions struggled with the effects of the accident; I want, in Book 2 of Chernobyl, to hear the voices of prominent writers and scientists ring out with their reflections on the future of humanity in the anxious light of the accident at the Chernobyl nuclear power station.

It is the readers' attention and support which inspires me to this.

That's it, the Zone!
And immediately such a chill over my skin...
Every time it's that chill, and even now I still don't know
if that's how the Zone receives me or
if it's the Stalker's little nerves playing tricks.

Reason is the ability to use the powers of the surrounding
world without ruining that world.
O. Strugatsky, B. Strugatsky,
Picnic by the Roadside, 1972.

1

Reflections

A year has passed since the accident at the Chernobyl power station.

A year, no more.

But how distant, how idyllically cloudless that pre-Chernobyl world now appears to us—calm, unhurried, self-confident, plunged for years as it were in somnolent, indulgent, all-permitting placidity.

For everyone who participated, directly or indirectly, in the tragedy of Chernobyl, time seems to have split into two unequal parts: before 26 April 1986 and after. One of the heroines of our narration, Aneliia Perkovska, has some apt words to say about this: 'It was really reminiscent of the war. Every one of us in the Town Committee has kept until now this sense of a boundary: before the war and after the war. We simply say: that was before the war.'

The time that has elapsed since the accident, particularly the first, most difficult months, months which seemed to last a whole eternity, can be divided into several epochs, stages, periods, moments—call them what you like—with their peculiar features and characteristics, with clearly outlined spans—lasting from the tragic Ukrainian spring of 1986, strangely lovely, in the snow-white blossom of the gardens and the full flood of the rivers, a spring which henceforth will enter all textbooks of history, all chronicles and legends of humanity, to the deep and dark autumn, when in Chernobyl a meeting took place: work was complete on the erection of the sarcophagus, the structure which covered over the ruins of Block No. 4.

A year, the blink of an eye in the history of humanity, a year isn't a particularly long span of time even in the life of any person. But in the course of that year—no, not a year, but just a few months—we all suddenly matured, grew up by a whole epoch, we became harder and more exacting both toward ourselves and toward those who take responsible decisions, those in whose hands human existence and the fate of nature rest; we began, in a different and more severe way, to evaluate the deeds and actions of all those months, the words pronounced and published during that time, so difficult a time for the people.

For the price which had, and would have, to be paid for Chernobyl was too high.

In the communication of the Central Committee of the CPSU and the Council of Ministers of the USSR of 14 December 1986, preliminary summaries are given of how, over a very compact period, large-scale tasks connected with removing the consequences of the accident at the Chernobyl nuclear power station were resolved, and certain figures and facts are included which give an idea of the extremely complex and unique work undertaken to save the destroyed power block, work which was carried out in difficult circumstances.

A full interpretation of what happened (let us remember the Great Fatherland War) is a matter for the future, perhaps the distant future. No writer or journalist, however well informed he might be, could do that today. The time will come, I firmly believe, when the Chernobyl epic (the thought never leaves me that this is indeed an epic, which, in its colossal scale, touches the fundamental questions of people's existence: of life and death, war and peace, the past and the future) will appear before us in all its tragic fullness, in all its polyphony, in the grateful biographies of the real heroes and the scornful characterizations of the criminals who allowed the accident and its grievous consequences—they must all be cited by name!—giving fine and precise figures and facts, giving the complexity of their everyday circumstances and official cunning, of people's hopes and illusions and giving the variety of moral positions taken by the participants in the epic. I think that, in order to create such an epic we will require new approaches, new literary forms, different, let us say, from *War and Peace* or *Quiet Flows the Don*. What will those approaches and forms be? I do not know.

And all the while... All the while I feel I want to propose to the reader my own original presentation of the documents and facts, of the testimony of witnesses—shortly after the accident it fell to my lot on several occasions to be in the Zone and in the places adjoining it.

The Chernobyl explosion took people into a new period in the development of civilization, a period about whose possibility only writers of science fiction had conjectured vaguely and intuitively. The majority of rational and optimistically inclined scientists and technically minded pragmatists, because of their limited imagination and consequent self-confidence, were incapable of foreseeing any such thing and, clearly, did not desire to. It is only individual, very far-seeing scientists who recently began to dwell on the catastrophic consequences of an incredible concentration of industrial and scientific forces. The words of Academician V.A. Legasov, published in the pages of this story, testify to this.

In the space of a few days we as it were took a step from one epoch, the pre-atomic epoch, into an unknown epoch which demands a fundamental restructuring of our thought. Not only human character, but also many of our conceptions and methods of work underwent severe examination.

Fate has given us the opportunity to peep over the edge of night, that night which will fall if nuclear warheads begin to explode. The Chernobyl accident has brought humanity a series of new problems, not only scientific and technical ones, but also psychological ones. It is very difficult for our consciousness to reconcile itself with that absurd situation where mortal danger does not even have taste, colour and smell, but is measured only by special apparatus, which at the time of the accident was, incidentally, either not available or not ready to work.

The accident showed us that man, if he wishes to survive, must develop a new, 'apparatus' way of thinking, complementing our sense organs and current methods of investigating our environment (for example, microscopy and chemical analyses) with Geiger counters.

The danger in Chernobyl and around there was diluted in the fragrant air, in the pink and white blossom of apple and apricot trees, in the cloud of dust on the streets and roads, in the water of village wells, in the cows' milk, in the fresh green of the gardens, in the whole of idyllic springtime nature. But really only springtime?

Already in autumn, when I was in the Polissia region chatting with people of the villages of Vilcha and Zelena Poliana, I became convinced of how the new demands of the atomic age were entering the consciousness and daily life and customs of people. The former, eternal cast of village life had come into conflict with the new realities of the post-Chernobyl world: dosimeter operators told me how very difficult, almost impossible it was to clear the thatched village roofs of radiation; burning leaves were very dangerous. We became convinced of this in Vilcha when we placed a dosimeter against a bonfire which had been lit in a farmyard by careless farmers: the device reacted with a considerable increase in the reading. So much for your 'and the smoke of the fatherland for us is sweet and pleasant.' And so, as a result, burning firewood was forbidden here; as one doctor aptly said, every stove in the Polissia region had been converted into a little fourth reactor. The population was provided with coal.

Who a year ago could have known that an increased radiation level would show itself in mushrooms, peatbogs, blackcurrant bushes and in villages at the corners of buildings where the rainwater ran from the roofs...

Since it was imperceptible, the danger aggravated some

people's sense of insecurity, but other people's most reckless disdain: more than one such bold person paid with his health for his 'boldness,' ignoring the simplest and, it must be noted, quite effective precautions.

It is only by an objective knowledge of the real situation, unwarped by anyone's 'optimistic' good will or concealed by layers of mystery, it is only by taking rational precautions and by a continuous control of radiation levels that people in the danger zone can have the indispensable feeling of security. That is one of the indisputable lessons of Chernobyl.

When I was in the emergency zone, observing how great a tragedy had unexpectedly befallen tens of thousands of people, I often recalled our literary discussions on a topical theme, on the present and future of the novel or story, on the positive hero and the need to 'study' (!) life and other things which at that time seemed very important to us. How academic and remote from this life they seemed there, in the Zone, where before my eyes an unprecedented drama had unfolded, where the essence of humanity was extremely quickly revealed, as in the war: all the masks suddenly flew off people's faces, like leaves from trees under the effect of defoliants, and the pompous-sounding prattlers, who at meetings exhorted us to 'acceleration' and 'activization of the human factor,' now turned out to be common cowards and scum; it was silent, unnoticed plodders who were the real heroes.

Let us just take the old fireman, 'grandad' Hryhorii Matviiovych Khmel and his story, unhurried just like village life. He and his two sons, also firemen, suffered somewhat during the power-station accident and were in various hospitals in Moscow and Kiev. His wife was evacuated from the village near Prypiat to the Borodianka area and carried on working, preparing food, which she took out to machine operators in the fields. What sort of literary or social problems, often quite trivial and wretched, that beset us in our lives, can be compared with the drama of these people, who conducted themselves with great human dignity? As I listened to the story of the thoughtful Ukrainian Khmel, I for some reason thought of Gogol's *Taras Bulba.* After what I learned and saw in Chernobyl, there was a time it seemed I would never take up my pen again: all traditional literary forms, all the subtleties of style and intricacies of composition—it all seemed to me infinitely remote from the truth, it all seemed artificial and useless. Several days before the accident I had finished my novel *Causes and Effects*, which tells of the doctors in a laboratory investigating particularly dangerous infections, who are struggling with an illness as deadly as rabies; and although some situations in the novel are by a strange coincidence similar to those I was to see (bearing in mind a difference of scale from what happened, of course), the novel somehow was very quickly obliterated from my consciousness, was pushed back somewhere, into a 'time of peace.'

It was all swallowed up by Chernobyl.

Like a gigantic magnet, it attracted me, it excited my imagination, it forced me to live in the Zone, its strange, twisted reality, to think only of the accident and its effects, of those struggling for their life in clinics, trying to tame the atomic genie in immediate proximity to the reactor. It seemed base and inconceivable to stand aside from events which were inflicting such calamity on my people. For long years before April 1986 I had been pursued by a feeling of guilt, guilt because I, a native of Kiev, a writer, a doctor, had passed by on the other side of the tragedy of my native town, the tragedy which had occurred at the beginning of the sixties: the damp sand and water accumulated in Babyi Iar, which the authorities wanted to make into a recreation area, broke through a dike and poured into Kurenivka, causing much destruction and human victims. For long years Ukrainian literature (and I along with it) had been silent about this catastrophe, and only recently Oles Honchar in his story *The Black Ravine* and Pavlo Zahrebelny in his novel *Southern Comfort* turned to the events of that terrible dawn in early spring... And why did I remain silent? I could have collected facts, the testimony of witnesses, I could have found out and named those guilty of this calamity... But I didn't. Very likely I had at that time not acquired the maturity to understand certain very simple, very important truths. And the time was such that my voice was not heard: it was weaker than a mosquito's whine: I had behind me at that time only my first publications in *Iunost* and *Literaturnaia gazeta*, and I was only just writing my first story *As in the War*... I say this not in justification, but for the sake of the truth.

Chernobyl is something I perceived quite differently, not only as my personal misfortune (nothing actually threatened me), but as the most important event in the history of my people since the Great Fatherland War. I could never have forgiven myself my silence. True, being at first a special correspondent of *Literaturnaia gazeta*, I saw my task in a rather narrow way: I would write about the doctors taking part in eliminating the effects of the accident. But the simple course of life forced me progressively to broaden my sphere of investigation and to meet hundreds of the most diverse people: firemen and academicians, doctors and policemen, teachers and power workers, ministers and soldiers, Komsomol workers and metropolitans, an American millionaire and Soviet students.

I listened to their stories, recorded their voices on tape, and then, deciphering these recordings by night, again and again I was struck by the genuineness and sincerity of their testimony, by the precision of the details, by the aptness of their reflections. Turning these recordings into text, I strove to preserve the structure of speech, the peculiarities of terminology or jargon, and the intonations of these people, resorting to editing only when absolutely necessary. It seemed very important to me to preserve the documentary and uncontrived character of these human confessions.

I wanted the truth to be preserved.

I am aware of the utter incompleteness of the material which I offer to the reader: the testimony of the witnesses which I cite here applies primarily to the first, most serious stage in the accident; there is certainly much to say about the construction of the sarcophagus and about the events connected with the decontamination of the locality, and about the extremely rapid construction of fifty-two new villages in the Kiev area, and about how the state offered compensation for the material losses of the victims and, of course, about the selfless work of the medical workers in the Zone and beyond it. How many extraordinarily interesting human destinies, how many unsung heroes! But I do not consider my work finished and shall continue to collect material in order to complete this story.

2

That Bitter Word 'Chernobyl'

Chernobyl.

A pleasant little provincial Ukrainian town, swathed in green, full of cherry and apple trees. In the summertime many people from Kiev, Moscow and Leningrad loved to holiday here. They came here for a long time, not infrequently for the entire summer, with their children and members of their households, they rented 'dachas,' in other words, rooms in wooden one-storeyed buildings, they prepared pickles and preserves for winter, picked mushrooms, which were to be found in abundance in the local woods, sunbathed on the blindingly clean sandy banks of the Kiev Sea, and fished. And it had seemed that here the beauty of Polissia nature had blended astonishingly harmoniously and inseparably with the four blocks of the power station, encased in concrete and situated not far to the north of Chernobyl.

So it had seemed...

When I arrived in Chernobyl at the beginning of May 1986, I (could I have been the only one?) peeped into the strange and incredible world beyond the looking glass, tinged with invisible and consequently even more ominous hues of heightened radioactivity. I saw something which the day before had still been difficult to imagine even in the most fantastic dreams, even though everything had an overall ordinary appearance. And later, on subsequent visits, everything already did seem normal...

But first...

It was a town without inhabitants, without the resonant voices of children, without the normal everyday, provincial unhurried life. The shutters were tightly closed, all the buildings, offices and shops were locked and sealed. On the balconies of the five-storeyed buildings near the fire station there stood bicycles and the washing was drying. No domestic animals remained in the town, the cows didn't moo in the morning, there were only wild dogs running around, hens clucking, and birds singing their carefree songs in the leaves of the trees. The birds did not know that the dusty leaves had, during those days, become a source of increased radiation.

But, even abandoned by its inhabitants, the town was not dead. It was alive, it struggled. But it lived according to the strict and for us completely new laws of the extraordinary conditions of the atomic age. In the town and around it there was a great concentration of technical equipment: powerful bulldozers and tractors, mobile cranes and earth-moving machines, excavators and concrete-carriers. Facing the Regional Committee of the Party, close by the monument to Lenin, an armoured troop carrier had stopped dead, and a young soldier in a gas mask was peering out of it. Under spotted camouflage nets radio stations and military freight trucks had been deployed. And in front of the Regional Committee and the Regional Executive Committee, from where the leadership was directing the whole operation, there stood dozens of cars: black Volgas and Chaikas, as if there was some high-level conference in progress. Some of these cars, which had 'contracted' radiation, had afterward to be left for ever in the Zone... At the approaches to Chernobyl there were numerous dosimeter checkpoints, where there was a very strict inspection of cars and tractors; in special areas soldiers wearing green anti-chemical protection suits decontaminated technical equipment which had come out of the Zone. Sprinkling machines continuously and liberally washed the streets of Chernobyl, and numerous traffic-regulation officials stood around, just like on the busy main roads of Kiev on the days before holidays.

But what is the history of this little town, destined to enter the chronicle of the twentieth century?

There lies on the table before me a small and—putting it as precisely as possible—comfortably and old-fashionedly published booklet which came out over one hundred years ago, in 1884. Its title is very engaging for the modern reader: *The Town of Chernobyl in Kiev Province, described by L.P. (Retired Soldier).*

With the scrupulousness of a real military man, living in leisure and not knowing what to do of use, the author studied the geography, history and economy of this unimportant little town, situated one hundred and twenty versts to the north of Kiev. 'Early historians relate,' writes L.P., 'that when the Great Prince of Kiev Mstyslav, son of Monomakh, in 1127 sent his brothers against the Kryvychy along four roads, Vsevolod Olgovych was ordered to proceed through Strezhiv to the town of Borysiv. Strezhiv was considered the most southerly little town in the Polatsk Principality, where Rohvold around 1160 settled Vsevolod Hlibovych. In the time of this prince Strezhiv, later named Chernobyl, was considered an apanage principality.'

'In 1193 in the chronicle Strezhiv is already called Chernobyl. It is written: 'The Prince of Vyshhorod and Turov, Rostyslav—the son of the Great Prince of Kiev Riurik (he ruled from 1180 to 1195), "rode hunting from Chernobyl to Tortsyiskyi".'

The author traces in detail the complex strands in the history of Chernobyl—who indeed did not own the town? At the end of the seventeenth century Chernobyl came into the possession of the Polish nobleman Chodkiewicz, and right up to the October Revolution the Chodkiewicz family owned over fifty thousand acres of land here.

The name of Chernobyl made a brief and tantalizing appearance in the history of the French Revolution: during the

Jacobin dictatorship: on 30 June 1794, a native of Chernobyl, the 26-year-old Polish beauty Rozalia Lubomirska-Chodkiewicz was guillotined in Paris, sentenced by the Revolutionary Tribunal, accused of links with Marie-Antoinette and other members of the royal family. Under the name 'Rozalia of Chernobyl' this blue-eyed blonde was immortalized in the writings of contemporaries...

Ancient Chernobyl gave its bitter name (*chernobyl* is the common wormwood) to the powerful nuclear power station whose construction began in 1971. In 1983 four power blocks, delivering four million kilowatts, were working. Many a person not only abroad, but even in our country and until now, after numerous publications and television programmes, had a more or less vague idea that Chernobyl, which had remained a rural regional centre, had, in the years preceding the accident, hardly had any contact with the nuclear power station. The 'capital' for the power workers was the young and rapidly growing town of Prypiat, eighteen kilometres north-west of Chernobyl.

In the album *Prypiat* (photographs and text by Iu. Ievsiukiv) brought out by the Kiev publishing house Mystetstvo in 1986 we read:

'It was called Prypiat after the full-flooded beauty of a river which, capriciously winding its way in a blue ribbon, unites Belorussian and Ukrainian Polissia and carries its waters to the grey Dnieper. The town owes its appearance to the construction here of the V.I. Lenin Chernobyl Nuclear Power Station.

'The first page in the chronicle of the biography of working Prypiat was written on 4 February 1970, when the first wooden picket was hammered in and the first scoop of earth was taken out. The average age of the inhabitants of the young town is TWENTY–SIX YEARS. Every year over a thousand babies are born here. It is only in Prypiat that you can see a pram parade, when of an evening the mothers and fathers walk out with their youngsters... Prypiat is confidently stepping out into the future. Industrial enterprises are increasing their productive capacities. In the next few years a power-industry technical college, another secondary school, a pioneers' palace, a young people's club, a business centre, a covered market, hotel, new coach and railway stations, a stomatological clinic, two-screen cinema, 'Children's World,' department store and other things will be built. The drive into the town will be adorned by an amusement park. According to the general plan Prypiat will have a population of around eighty thousand. The Polissia atom-city will be one of the most beautiful towns in Ukraine.'

This colourful album was presented to me in the empty main administrative building of Prypiat, its 'White House,' by Aleksandr Iurievich Esaulov, the deputy president of the Prypiat Town Executive Committee and one of the heroes of our story. I walked with him along dead corridors, looked into empty offices: furniture pushed aside, papers scattered on the floor, safes unlocked, piles of empty Pepsi-Cola bottles in the places where the Government Commission held its meetings (from off the doors I took as souvenirs the hastily inscribed slips of paper—who worked where), newspaper files opened at 25 April, withered flowers in vases... And over it all the overpowering smell of disinfectant spread to stop rats breeding.

That day Esaulov and I were the only inhabitants of this forsaken handsome town. There were only we two and a few security policemen, guarding the buildings left by the townsfolk. The drive into the town was adorned not by an amusement park but by a fine-mesh fence made from barbed wire, rigged with an alarm, so that uninvited looters might not take a fancy to getting through, into the Zone, to extract some profit from the radioactive things left behind in thousands of apartments. There were such people.

3

Before the Accident

Precisely a month before the accident, on 27 March 1986, in the newspaper *Literaturna Ukraina,* the organ of the Union of Ukrainian Writers, there appeared L. Kovalevska's article 'Not a Private Affair.' It should be recalled that for several years the newspaper had already had a permanent column 'The *Literaturna Ukraina* Eye on the Chernobyl Power Station,' clarifying the various events in the life of the power station. This article, which was fated to create such a sensation all over the world (after Chernobyl the western mass media vied with each other to quote it), at first attracted no attention: at that time Kiev writers were getting ready for their general meeting and most of them were far more interested in the coming personnel changes in the organization than in the affairs of the power station.

L. Kovalevska's article had no relevance to the operation of the fourth block of the Chernobyl power station; but many people, hearing of her article through rumours, have remained until now convinced of the opposite. The author concentrated the fire of her criticism—very professional and uncompromising criticism—on the construction of the fifth block, whose completion target had been reduced from three to two years. L. Kovalevska cited blatant facts of irresponsibility and shoddy work: for example, in 1985 suppliers had fallen short by 2,358 tonnes of metal components. And what they supplied had most often been defective... Furthermore, 326 tonnes of defective fine-mesh sheathing for the spent nuclear fuel depository came from the Volga Metal Components Works. And around 220 tonnes of defective pillars were sent for constructing the depository from the Kashin Metal Components Works.

'It's quite inadmissible to work like that!' was how L. Kovalevska ended her article. 'The prompt inauguration of the next power block is not the private affair of the builders of the Chernobyl power station. After all, "acceleration" is our activity too, our initiative, our perseverance, our awareness, and our attitude to everything which is done in our country.'

Quite honestly, when I read this article (and I read it, like many others, after the accident), it appeared to me that an experienced engineer had written it, some grey woman in eyeglasses, an expert in all those insipid construction terms and norms. How surprised I was when Liubov Kovalevska turned out to be a young woman, a journalist on the Prypiat newspaper *Trybuna enerhetyka*, and a talented poetess.

She has striking eyes—clear, with severe dots for pupils; now and then it seems that her expression reaches out somewhere far away, perhaps into the past or into the future, and then her expression is very sad. Her voice is a little hoarse, she smokes heavily.

And so, **Liubov Oleksandrivna Kovalevska**:

'I was accused of everything under the sun after the appearance of that article in *Literaturna Ukraina:* that I was incompetent, half-educated (the expressions they chose were, it's true, milder, but that's what they amounted to), that I had washed our dirty linen in public, that I was writing to Kiev newspapers to make my name.

'It's only when something incredible happens that people will believe and understand.

'In our newspaper *Trybuna enerhetyka* we wrote primarily about construction problems, but the Prypiat Town Committee of the party wanted us to do the impossible, to write about everything, about the town—after all, this was the only newspaper in the town. But there were three of us, we didn't have our own transport, and was it really possible for three poor women to run around that gigantic construction site? And not just run around, but then return to the office. Heaven help us if someone phoned and there was no one there, that would mean we weren't working.

'At first I was the editor, but when the conflict heated up I became a correspondent again; on the Town Committee they breathed a sigh of relief. You see, I always stood up for the newspaper's right to independence of thought, analysis, arguments and conclusions.

'I wrote the article for *Literaturna Ukraina* in one evening.'

'Tell me, was this a case of a journalist struggling for the truth being harassed by those in charge?'

'As far as the construction work was concerned it would be unfair to say that. But as regards the Town Committee or the power station administration, yes, it would have been like that. I didn't go to the Town Committee, didn't find out their opinion of the article, but rumours spread. Reliable ones. I found out they were going to summon me to the office. They could expel me from the party.

'But then the accident occurred...

'I consider that one of the reasons for the Chernobyl power station accident was the abnormal situation which developed there. A "casual" person could not get in there. Even an exceptionally intelligent person or a first-class specialist. You see, in the management there were whole dynasties, nepotism flourished. The wages were high, they got them through unhealthy conditions of work and this was done via a 'dirty network.' Even the workers wrote about the utter nepotism there. They were friends, acquaintances. If you criticized one, they would all rush to his defence without trying to get to the bottom of the affair.

'When an ordinary worker makes a mistake, he is punished. But when it's the administration, the people at the top, then they get away with it. It got to the point where the administration could avoid exchanging greetings with the workers, could talk to them in a condescending manner, could wrong them and offend them. Ambitions grew out of all proportion. It was like a state within a state. They didn't take into account that people couldn't help but see all this... And people came to our editorial office and begged us: "Just don't name me, you understand, I'll be kicked out of work, they'll destroy me, but you, you journalists, you can write about this." Could we refuse? A journalist hasn't the right to be a coward. But you couldn't do anything because it was impossible to name the person.

'When I was editor, I didn't take material to get approval. I don't know whether that was right or not, I just didn't do it. I'm the one responsible for my articles. I bear the responsibility myself, both as a communist and as a journalist.

'I conceived the idea of a series of articles for *Literaturna Ukraina.* The first one was to be about problems connected with the construction work. And the second, well, the second just had to be about the people running the place. About the moral climate in the Chernobyl power station. Let's be honest about this: the best of the construction workers went into management. For the sake of the wages. The management even made overtures to the good specialists. If you make an experienced construction worker a supervisor, then he knows the construction side from 'A' to 'Z.' He's a valuable member of staff. A supervisor is the one who checks how the building is going. But at that time there was a shortage of money for the construction work. Like it or not, many a worker let his standards fall, even the qualified construction workers. But how did these construction workers cope when it came to eliminating the accident? In the newspaper I read that they achieved their annual plan in one month. They're invaluable, those people. They can work, and they want to.

'So, lots of people went into management. And later they came to me in the editorial office and complained: "My goodness, how honestly people work on the construction site and what a difficult moral climate there is in the power station. It's as if you taken someone else's place... It's careerism, fighting for a job, a position".'

'And did they earn high wages?'

'Of course. Three hundred roubles and more. They always over-achieved the plan, "progress"... And if they were in a "dirty" zone, then there would be coupons for food here, rations there, there would be places on holidays, all the privileges you for some reason, I don't know why, don't get on the construction site. And if you were at the power station you received an apartment more quickly than if you were at the construction site, even though it's the construction workers who do the building; I suppose the power station was "important," and the ratio came out (I don't remember exactly) around seventy per cent to the management and thirty to the construction workers.'

'What would have been the main problem you touched on in your article? What would you have said in it?'

'I would have said that people have a duty to believe. I'm a subordinate and have a duty to believe my editor. To believe. So that I feel at one with him. To believe in his authority and his qualifications. In his competence. It's the same with the workers. If the people in the administration are honest, decent and guided by principle, then, naturally, the workers strive to be like them. But if one person has everything permitted him, and another one nothing, then envy rears its head, a psychological discomfort. People wonder: "What's the point of living the life of a fool, when next door people are living well, just as they like. They exhort us to honest work and enthusiasm, while all the while they themselves... acquire Czech lavatory pans from the hotel. The leave their own ones there and take the Czech ones." You see, this town is small, and the slightest mistake, the slightest moral flaw in a manager gets known very quickly. And all this gets discussed, turned over and over, rumours and gossip spread around, all the more as the suppression of criticism was rigorous.

'Seen from the outside the discipline seemed ok. Everyone was afraid, really afraid, to go off home first. But off he went if no one saw him. He was afraid of arriving late, but he would arrive late if no one saw him. There was a loosening of people's strong inner core, it all became rickety. And so, when the accident occurred, it turned out that it wasn't just the management which was at fault, but also those operators who...

'And so that article I thought of would have shown the connection between discipline and the violation of the elementary rules of security. Just imagine, you could see a man sitting at the control panel. There where you have the buttons and levers.'

'How do you mean, sitting at the control panel?'

'Well, this fellow just went and sat there. He can go and sit at the control panel. Just like that. And what do people say? "It operated in such a way that the systems duplicated each other. They protected people." Everyone believed in the systems. But they didn't protect people. And they didn't protect people because we got to the state—we, not just anyone, I don't even blame the administration, but just us, the people—we got to the state where we split into two. One half said it's necessary to do things one way, it's necessary to work honestly, while the second wonders: "But what if the other person doesn't do it?"

'At the Chernobyl nuclear power station centre there was a nuclear power engineering commission. I was at its meetings and often visited the power station. There wasn't a permanent pass even for our newspaper. The management wouldn't give one for fear that, God forbid, you wrote critical articles. But if you had the intention to write something good, then they'd show you everything. You just had to say in advance to the Party Committee where you wanted to go and with what objective.

'There were stops there caused by the personnel too: "whistling" in the steam pipes. How twisted the psychology of our personnel is! If a foreign delegation comes, then they're afraid. They understand that it's not possible to do things like that. Among ourselves we had the attitude, "It whistles, well, let it whistle!"

'After the article people said that I had prophesied the accident. I didn't prophesy anything, God preserve me from being a Cassandra, a prophetess of such calamities... But deep down, if I'm honest, I did always fear this. I wasn't easy. I was afraid because people said one thing, but really the situation was quite different. People talked about this, they talked about this in the safety section. When did I start to be afraid? People once came to me and brought documents, they showed me facts and figures, and what not, things I simply did not know, but at the time I just hadn't the courage to write about it. I knew it had no chance at all of being published.

'And I was afraid. And all the time I wanted to leave, honestly, to take my child away from here. I have a ten-year-old daughter and she is not in the best of health.'

4

The Accident

In May 1986 in one Kiev hospital I made the acquaintance of the following young people: Serhii Mykolaiovych Hazin, a twenty-eight-year-old senior turbine engineer, Mykola Serhiiovych Bondarenko, twenty-nine years old, an air-fractioning worker in the nitrogen-oxygen station, and Iurii Iurievich Badaev, a thirty-four-year-old engineer. They were united only by the fact that they were in the one ward, gradually emerging from that grievous condition in which they had been when they entered the clinic, and by something more essential, which had separated their lives into two parts: before and after the accident. That fateful night they had been working at the power station, in the immediate proximity of the affected block.

Calmly and clearly they recounted how it had all happened, how two strong shocks had shaken the station building, how the light had 'cut out' and everything had collapsed in clouds of dust and steam. The hall, where the block's control panel was located—the control centre of the whole power block—was lit up only by the sparking of short circuits. Only now and then their outwardly calm story was buried by a deep sigh or a painful pause, when recollections of that night re-emerged.

Iurii Iurievich Badaev that night had been working at

the information processing complex SKALA:

'SKALA is the brain, the eyes and the ears of the station. A computer does essential operations and calculations and presents it all to the block's control panel. If SKALA stops, they're like blind kittens.

'My position there is as an "electrician-fitter." Surprising? But that's how it is. By training I'm an electronic engineer. Usually in computer centres you have electronic engineers, but among us for some reason we're called "electrician-fitters."

'What happened was very simple. There was an explosion. I was on shift forty metres from the reactor. We knew there were experiments going on. The experiments were according to a previously planned programme and we were following this programme. Our computer registers all deviations and records them on a special tape. We were watching over how the reactor was working. Everything was fine. Then a signal came which meant that the senior reactor engineer had pressed the button to switch the reactor totally off.

'Literally fifteen seconds later there was a sudden shock, and a few seconds later a stronger shock. The light went out and our machine cut out. But some sort of emergency supply came on and from that moment we tried to save the equipment, because everyone needs our information. Moreover, this is the most important thing, this is the diagnosis of the development of the accident. As soon as we had the emergency supply, we started the struggle to keep our machine going.

'Immediately after the explosion we felt nothing at all. The fact is that hothouse conditions are created for our computer; a temperature of 22°C-25°C is maintained, with a constant pressure ventilation. We managed to get the machine going and to protect the "racks" (i.e., the computer—Iu.S.) from water, which by then was beginning to pour through the ceiling. The machine was working, and the diagnostic system was continuing. It was difficult to understand what it was registering. It was only then that we asked ourselves: what on earth has happened? We needed to take a look. And when we opened the doors, we could see nothing but steam and dust and the like. But then somewhere the "racks" which controlled the reactor were shut off. Well, where we were is the holy of holies; we have to do everything possible to maintain the monitoring. And I had to go to the twenty-seventh level, where the "racks" were. A level, that's a sort of storey. I rushed off along the usual route, but it was impossible to get to the level. The elevator was crumpled, crushed tight shut, and there were blocks of reinforced concrete on the steps and some sort of tubs; but the main thing was that there was no light. We still didn't know the scale of the accident, nothing at all. Nonetheless, I wanted to get there and even ran off for a flashlight. And when I came back with the flashlight, I realized that I wouldn't get through... Water was pouring from the ninth floor, it really was pouring. We took emergency shields and covered our computers, to protect them, so that SKALA would continue to work.

'Then we learned the extent of the accident. I really had to convince myself of this. Literally a few minutes before the accident Shashenok had called on us. He was one of those two fellows who died. We chatted with him just as we are talking with you; he had come to get something clear: "Do you have a direct link to the room on the twenty-fourth level?" We said that we had. They had to do some work there and after all, this was one of the colleagues who were carrying out the programme of experiments, noting down the reactor's performance. They had their own apparatus in that room. He said: "Lads, if I need a line, I'll come through you." "Fine," we said.

'And when we had already saved our equipment, a call came from that room where Shashenok had been working. A continuous call. We took the receiver: no one answered. As it later turned out, he couldn't reply. He had been crushed: he had broken ribs and his spine was twisted. Nevertheless, I tried to break through to him, I thought that he perhaps needed help. But they had already got him out. I saw them carrying him out on a stretcher.'

And the town slept.

It was a warm April night, one of the best nights in the year, when the leaves were just appearing in a green haze on the trees.

The town of Prypiat was sleeping, Ukraine was sleeping, the whole country was sleeping, still unaware of the massive misfortune which had come upon our land.

5

The Entire Guard Followed Pravyk

Firemen heard the first alarm signals.

Leonid Petrovich Teliatnikov, Hero of the Soviet Union, thirty-six-years old, chief of militarized fire brigade No. 2 of the Chernobyl power station, internal service major (now lieutenant-colonel):

'In Lieutenant Pravyk's guard there were seventeen people. That night he was on duty. If we are to speak about this guard as a whole, in contradistinction to what they write in the newspapers, then the third guard was not so ideal. And if it hadn't been for this incident, of course, no one would ever have written about it. It was a very original guard. It was a guard made up of personalities, one might say. Because each member was his own man. There were a lot of veterans there, a lot of original fellows.

'Volodia Pravyk, I think, was the youngest, he was twenty-four. By nature he was kind and gentle, and they sometimes let him down. He never refused anyone who made any kind of request. He considered that he had to make concessions. There was, perhaps, a certain weakness here on his part—there were conflicts, and he ended up at fault, because there were violations in the guard... but he kept to his line.

'He was a great enthusiast, Volodia Pravyk. He was a radio amateur, and a photographer. He was one of our active workers, the chief of staff of the Komsomol 'searchlight squad.' The 'searchlight squad' was probably the most effective means of struggling with shortcomings, and it came down hard on even the slightest ones. He wrote poetry too, and drew, and did this work with pleasure. His wife was a great help to him. They complemented each other perfectly. His wife graduated from a music academy and taught music in a kindergarten. Even to look at they were alike, both gentle, and their outlook on life and their attitude to work—all this was closely intertwined, was as one. A month before the accident their daughter had been born. Recently he had asked to be made inspector, and everyone agreed, but there was simply no one to replace him...

'Probably the most senior in the guard in age and length of service was Ivan Oleksiiovych Butrymenko. He was forty-two. He was one of those people who hold everything together. Everyone measured himself by him. Even the chief of the guard and the secretaries of the Party and Komsomol organization. Ivan Oleksiiovych was a representative on the Town Council, and as such did a great deal of work...

'In our brigade there also worked the three Shavrei brothers, Belorussians. The youngest, Petro, worked as a brigade inspector, and Leonid, the eldest, and Ivan, the one in between, worked in the third guard. Leonid was thirty-five, Ivan was two or three years younger, and Petro was thirty. Their attitude to work was: if you have to, you do it.

'Is it like that in life? Until you crack the whip, no one even moves. It's not only here that it's like that, but everywhere. In one's business, in one's teaching—one person tries to keep out of the limelight, to take a little rest, to take on some lighter work... There was none of that here. When the accident occurred, in spite of whatever discord there was in the guard, in spite of everything, the whole guard followed Pravyk, followed him without looking back... There bitumen was burning. The roof of the machine room was burning, and in terms of roubles the most valuable thing was the machine room.

'Everyone felt the tension, sensed the responsibility. I only had to call a name and up he'd run: "Understood!" And even without listening to the end, because he'd understood what he had to do. He was only waiting for the order.

'And no one held back. They sensed the danger, but they all understood: it had to be done. I only had to say the word, and they'd relieve one another. At a run. What was it like before the accident? "Why am I going? Well, why?" But now neither a word nor a murmur, and everything really was done at a run. That really was the main thing. Otherwise the fire would have taken far longer to extinguish and the consequences would have been far worse.

'When the fire began, I was on leave. I had thirty-eight days' leave. I was phoned at night, the transport controller phoned me. There was no transport, all the cars were out. I phoned the town militia division which was on duty, explained the situation, this and that, they always have cars. I said: "There's a fire at the station, on the machine room roof, please help me get one." He asked me my address again and said: "There'll be a car immediately."

'Up on top the roof was burning in one place, a second, and a third. When I got up there, I saw that it was burning in five places on the third block. At that time I still did not know that the third block was working; but since the block was burning, the fire had to be put out. This did not require any great firefighting effort. I took a look in the machine room: there were no traces of the fire. Fine. In the "stack" on the tenth level, where the central block control panel was situated, there was no fire. But what about the cable rooms? For us that was the most important thing. We had to go all the way round and get a good look. So all this time I was running around, I examined the fifth and eighth levels, I looked at the tenth, and at the same time clarified some things with the deputy chief engineer and the operational staff: what was more important, what and how... They said: "Yes, indeed, you have to put out the fire on the roof, because the third block is still working, but if the roof collapses and just one slab of concrete falls on a working reactor, there could be an additional depressurization." I had to know all this, because there were many places, the station was very big, and I needed to get everywhere.

'At the time I didn't manage to speak to Pravyk. It was only when he was being sent to the hospital, and only for that second literally a few words. At 0225 he had already been taken off to hospital. They had spent 15-20 minutes up there.

'Somewhere around 0345 I began to feel bad. I lit a cigarette, kept panting as before, and had a persistent cough. My legs felt weak, I wanted to have a brief sit down... There was nowhere to sit. We drove off to look around our positions and I showed them where to park the engines. We went to the director, I needed to telephone, to report on the situation. But at the station there was nowhere to phone from. Many rooms were closed, there was no one there, and though the director had several telephones, they were all being used. The director was speaking. Just then he was going off full blast on his telephones. We couldn't phone from there. So we went out to the brigade.'

And the alarm was growing.

At this point one important detail has to be clarified. Apart from the guard of Lieutenant Pravyk of Militarized Fire Brigade No. 2, the guard of Lieutenant V. Kibenok of Independent Militarized Fire Brigade No. 6 had also been alerted. Few people know that V. Kibenok's guard belonged to quite another sub-department of the fire service—Brigade No. 6, located in Prypiat. Even now the sub-department stands—it's a little building on the outskirts of Prypiat—and behind the glass doors the powerful fire engine has fallen silent for ever, as a monument to the achievement of Kibenok's guard.

L. Teliatnikov: 'Our Brigade No. 2, and in it V. Pravyk's guard, protected the nuclear power station. It was a site station. The town brigade, in which V. Kibenok worked, looked after the town. They learned of the fire at once. In the event of a fire we automatically receive a high call number, and this is immediately passed on to the central fire-brigade liaison unit. It is communicated by radio or telephone via the Town Brigade. In relation to us the Town Brigade is the main one. So, once they know that a fire has broken out, they automatically know that they have to turn out.

'As I have noted, Pravyk's guard was at first in the machine room. They got the fire out there, and the unit was left on duty under his direction, because the machine room was still at risk. And the Town Brigade, when it arrived a little later, was sent to the reactor section. At first the machine room had been the focal point, and then the reactor section. And so Pravyk then even went and left his guard and offered his help to the Town Brigade. Out of our guard only Pravyk died. The other five men who died were fellows from Town Brigade No. 6. It turned out that they were the first to start putting the fire out in the reactor. It was there that it was the most dangerous. From the point of view of the danger from radiation, of course. But from the point of view of the danger from fire, it was in the machine room, because it was there that our guard operated at the start of the accident.'

And the alarm was growing.

At the very outset of the accident V. Pravyk gave the alarm signal to all the Kiev area fire brigades. In response to this signal the fire sub-brigades of neighbouring populated areas were sent to the nuclear power station. The reserve was placed on emergency stand-by.

Hryhorii Matviiovych Khmel, fifty years old, fire engine driver of the Chernobyl regional fire brigade:

'I like playing draughts. That night I was on duty. I was playing with our chauffeur. I said to him: "No, Myshko, you're doing it wrong, you're making mistakes." He was losing. Our chat stretched out till somewhere around midnight, then I said: "Myshko, I think I'll be going to bed." And he said: "I'll go off for a walk with Borys." "Fine, do."

'We have trestle beds there, and I set one up, took a mattress and a clean quilt, a quilt from the little cupboard, and I put it under my head and lay down. I don't know whether I dozed for a long time or not, but then I heard something: "Yes, yes, we're on our way!" I opened my eyes wide and saw Myshko, Borys and Hryts standing there. "Let's go!" "Where?" "Volodia will get the information this instant." Then as soon as he had got the information, the siren whined. The alarm. "Where to?" "To the Chernobyl power station."

'Myshko Holovnenko, the driver, set off without a pause, and I set out second. There are two engines in our brigade. I was in one and he in the other. Well, you know, if often happens that when we set out we don't close the doors, and the doors are glass, and often the wind breaks them. So, normally, the one who comes out last must close them. Pryshchepa and I closed the garage. I thought to myself: I'll catch up with Myshko, I've got a ZIL-30. So off we went, and I chased him at around 80 kilometres an hour. Above my head the walkie-talkie was crackling, he was calling Ivankiv and Poliske out, the traffic controller was calling us out. I could hear that this was a real alarm call. I thought that it was really something serious...

'Then I caught up with Holovnenko's engine just before the power station, so that there'd be no confusion over the building, so that it'd be two engines together at once. When I caught up with him, I stuck with him. We drove up. As soon as we were there, at the power station administration building, we could immediately see the flames. It was like a cloud, with red flames. It struck me there was going to be work to do. Pryshchepa said: "Well, grandad Hrysho, there'll be lots of work." We arrived there ten or fifteen minutes before two o'clock. We looked—there were none of our engines, neither from Brigade No. 2 nor from Brigade No. 6. What did this mean? It turned out they'd parked on the north side of the block. We got out—where now, then? We surveyed the area. Graphite was scattered all around. Myshko said: "Graphite, what's all this?" I kicked it away. But a fireman on the other engine picked it up. "It's hot," he said. Graphite. There were various pieces, big and small, pieces you could take in your hands. They had fallen all over the road and we were all tramping over them. Then I saw Pravyk run up, the lieutenant who died. I knew him. I worked two years with him. And my son, Petro Hryhorovych Khmel, is a guard chief, just like Pravyk was. Pravyk was chief of No. 3 guard, and Petro of No. 1. They finished their education together... And my second son, Ivan Khmel, is the fire-brigade inspection chief for the Chernobyl region.

'Yes, so Pravyk ran up. I asked him: "What now, Volodia?" And he said: "Let's get the engines on the 'sukhotruby.' Over here." And now the engines from brigades No. 2 and No. 6 came up and turned toward us. We had two and they had three tank-trucks and a mechanical ladder. Five engines on this side. All together! We moved the engines up to the wall onto the 'sukhotruby'. Do you know what 'sukhotruby' are? No! They're empty hoses to which you have to connect the water and then drag them up there, onto the roof; and we had to take hoses, and join the hoses and put out the flames.

'We set the engine on the 'sukhotruby,' really close up, we had a very powerful engine, Petro Pyvovar's from No. 6 on the hydrant next to it; there's no longer any room for mine there. Then Borys and Mykola Titenok and I said: "Let's have a hydrant!" Well, they found us one, and that was difficult at night. During training we knew that our hydrant was on that side according to the plan, but it turned out that it was on this side. We found the hydrant, drove the engine up close, quickly threw the hose out, and they—Titenok and, I think, Borys, I can't remember—rushed forward to Myshko's engine. Three hoses twenty metres each, that makes sixty metres.

'We didn't have much idea about radiation. Whoever was working didn't have any idea. The engines delivered the water, Myshko filled the tank with water, the water went up, and then those lads who died went up, Mykola Vashchuk, and others, and Volodia Pravyk himself. At the time it was only Kibenok I didn't see. They scrambled up there using a step-ladder. I helped them set it up, it was all done very quickly, all this was done, and I didn't see them again.

'So, to work. We could see the flames, they were burning hot, and making such a cloud. Then there was a chimney there—I don't know the layout, something square—and further off from there it was burning too. When we looked, it wasn't flames burning, but sparks were already beginning to fly. I said: "Lads, this is already going out."

'Leonenko, the deputy chief of Brigade No. 2, came up. Well, we knew that Pravyk and Teliatnikov had already been taken away, so we began to understand what radiation was. They told to us to come into the refectory and take some powders. As soon as I came in, I asked: "Isn't Petro around?" Petro was due to replace Pravyk at eight in the morning. They said: "No. " But he's been put on alert. Just as I went out, the lads said: "Grandad, Petro Khmel's been taken there as substitute. I thought: that's that, a screw-up.

'At this point lots of engines drove up, and our bosses came. A Volga came from the administration, and cars arrived and drove up from Rozvazhiv and Dymer. I saw Iakubchyk from Dymer, we know each other, he's a driver. "Is that you, Pavlo?" "Yes, it's me." We got into the engine and drove off to the first building; there we were taken into into a room and checked for radiation. Everyone walked up and he wrote: "Dirty, dirty, dirty, dirty." And they just didn't say anything. It was at night. They took us off to a shower room, to get washed. They said: "Get undressed ten at a time, and leave your clothes here."

'I had leather boots and pants, no waterproof because I was a driver, and an underjacket, robe and protective shirt. He said: Take your documents with you, and your keys, and go straight into the shower." Fine. We got washed, went out through another door, and there they gave us clothes and slippers. It was all serious. I went out into the street and I could see, it was all visible and clear, I could see. My Petro was coming along, in his uniform, in a cape, with his fireman's belt, cap and leather boots. "It's you, son, are you going there too?" At this people thundered toward him, because he was going where we had got washed; they grabbed him and led him off, not letting him in, to put it briefly. He just said: "You here, dad?," and off they took him.

'Then we were taken to a civil defence cellar. There it was

quiet and there were beds, it was seven o'clock. Then it was eight o'clock, and nine o'clock, when I saw Petro coming. He'd changed clothes. Well, my son came, he sat down, chatted with me, ordinary domestic questions, nothing about this at all. He said: "I don't know what to do at home, Dad, I feel sort of sick." Then: "Dad, I think I'll go to the medical unit." "Go on, then." And he went.

'I didn't even ask him what he'd been doing on the roof, there hadn't been time to ask.

'And Ivan, my second son, had also been called out by the alarm. He belonged to the Chernobyl regional section. They had got him up around 6am. They had sent him on watch or somewhere. He had a little truck Urals car, and he just drove around.

'At first it was as if I felt nothing. Then, at first, not sleeping, then we got agitated, then I got scared, I shook so much, you know, all this . . . '

Hero of the Soviet Union Lieutenant Volodymyr Pavlovych Pravyk.

Hero of the Soviet Union Lieutenant Viktor Mykolaiovych Kibenok.

Sergeant Mykola Vasyliovych Vashchuk.

Senior Sergeant Vasyl Ivanovych Ihnatenko.

Senior Sergeant Mykola Ivanovych Titenok.

Sergeant Volodymyr Ivanovych Tyshchura.

Six portraits in black frames, six fine young men look at us from the wall of the Chernobyl fire station; their expressions seem to be sorrowful, and bitterness, reproach and the dumb question: "How could this happen?" have become fixed in them. It's what we think we see. But that April night, in the chaos and alarm of the fire, their expressions held no sorrow or reproach. There wasn't the time. They were working. They saved the nuclear power station, they saved Prypiat, Chernobyl, Kiev, all of us.

It was June 1986 when I came here, to the holy of holies of the Internal Affairs Ministry of the UkrSSR—the Chernobyl Fire Station—which had become the centre for all the work in the Zone to control the fire. An extraordinaily scorching

June, with the sun shining fiercely in the sky and not the slightest hint of a cloud. And all this happened not by God's will but by man's: pilots were pitilessly destroying the clouds in the zone of the nuclear power station, using special methods to control the sky from the air.

The fire station is a fine building, almost like a dacha. I looked at those doors which 'grandad' Khmel closed after him as he set off to the fire. The glass hadn't broken. Two firemen with hoses in their hands were washing down the yard, over which a sultry wave of hot air was slowly rising. Ready to set out, cleaned until they shone, the red-and-white fire engines with the Cherkasy, Dnipropetrovsk and Poltava region registrations stood there. Near the reactor there was a round-the-clock rotation of firemen: you could expect anything. Apart from that, the united fire detachment on duty during those hot days in Chernobyl was fated to participate in the struggle with 'ordinary' fires, which were no rare event in the local wooded and swampy places, especially in years of drought: in the Zone the peatbogs were burning. And, like everything in the Zone, these 'ordinary' fires were also extraordinary: together with the smoke radioactive aerosols flew up into the air, something which just could not be allowed...

Here I met the chief of the administration of the UkrSSR Internal Affairs Ministry's fire defence Major-General Pylyp Mykolaiovych Desiatnykiv and the chief of the firemen's detachment Colonel Evgenii Iukhimovich Kiriukhantsev. Colonel Kiriukhantsev is a Muscovite, the standard type of military intellectual: disciplined, handsome and precise. He told me that at the beginning of June in their brigade there took place a most strange, but very significant 'friendly' trial. The trial was about the fact that two firemen... 'pinched' two roentgens more that they had the 'right' to, during a real operation (all operations before being carried out were carefully planned and more than once rehearsed with a chronometer).

Just think!

In May they would probably have praised them and

declared them heroes. In June they punished them. The times changed so swiftly in the Zone, and the very attitude to this capacious concept 'heroism' changed.

Only the attitude to the men whose portraits hung in black frames on the wall of the Chernobyl fire station has not changed and never will change.

6

Bilokin From The Ambulance Service

Valentyn Petrovych Bilokin, twenty-eight years old, doctor in the emergency medical unit of the town of Prypiat.

'On 25 April I came on duty. In Prypiat there is one emergency brigade: a doctor and a medical assistant. And altogether we have only six ambulances.

'As there were a lot of calls, we split up: the medical assistant looked after the calls to 'chronics'—if injections were needed—and the doctor after the complex cases and children. During this duty period we worked separately, like two brigades: the medical assistant Sasha Skachok, and I. Masnetsova was the controller. From eight o'clock that evening everything seemed to take off, to hurtle with amazing rapidity. No, at first everything was calm at the nuclear power station, but not calm in the town. I was driving around all the time, not even getting out of the car. At first there was some drunkenness, someone went and threw himself out of a window, no, he didn't kill himself, he was fine, just "totally soused"... Then there were calls for children and we went to one old dear, and then sometime in the evening, just before midnight, I could remember it well as the night was very chaotic, there came a call: a thirteen-year-old boy with bronchial asthma, the attack had become serious. And it had become serious because a neighbour phoned and didn't give the number of the apartment. I drove out onto Construction Workers' Avenue, but it was already midnight and the house was big. I looked, walked around and around—no one. What could I do? I couldn't wake everyone up. So off I went.

'When I got back, Masnetsova said: "They've phoned with the number of the apartment. Back I went and, on arriving, the neighbour began to swear at me because I had arrived late. I said: "Ok, but I didn't know the number." And he retorted: "You ought to have." But I honestly didn't know, it was the first time I'd been to that boy. In the house the neighbour shoved me around and was close to getting into a fight, so I took the boy down to the Riga ambulance and gave him an intravenous injection of euphilline. And the neighbour was still threatening to lay a complaint against me...

'So when we were returning to the hospital—I was with the driver Anatolii Gumarov, an Ossete, about thirty years old, we saw that. What was it like? We were driving along at night, the town was empty, asleep, and I was sitting next to the driver. I saw two flashes to the side of Prypiat; at first we didn't realize that they came from the nuclear power station. We were driving along Kurchativ Street when we saw the flashes. We thought they were shooting stars. Since there were buildings all around, we could not see the nuclear power station. Just flashes. Like lightning, perhaps, a little bigger than lightning. We didn't hear any thunder. The motor was running. Later at the block they told us that it had really boomed. And our controller heard the explosion. One, then a second one straight afterward. And Anatolii said: "Shooting stars or not shooting stars, I just don't understand." He is a hunter himself, so it astonished him somewhat. The night was peaceful and starry, nothing special... When we arrived at the medical unit, the controller said that there had been a call. We arrived at 0135. The call had come to go to the power station, and the medical assistant Sasha Skachok had gone off there. I asked the controller: "Who phoned, what sort of a fire is it?" She didn't say anything to the point, whether I needed to go or not. Well, I decided to wait for information from Sasha. At 0140-0142 Sasha phoned me and said that it was a fire, with people burned, and that a doctor was needed. He was very excited, gave no details, and hung up. I took my bag and drugs because of the burns, and told the controller to get in touch with the medical unit chief. I took two empty ambulances and myself went with Gumarov.

'Going straight to the power station in a Riga ambulance takes around seven to ten minutes.

'We left town by the Kiev road, then turned left to the power station. It's there that I met Sasha Skachok; he was coming to meet us in the medical unit, but their emergency light was on, so I did not stop them since, if that light is on, then it's something serious. We carried on to the power station.

'The guard was standing at the gates and asked us: "Where are you going?" "To the fire." "Why haven't you any special clothing on?" "But how was I to have known that special clothing was necessary?" I had no information. I was wearing my doctor's smock, it was an April evening, a warm night, I didn't have a cap, nothing. We drove in and I met Kibenok.

'During my conversation with Kibenok I asked him: "Are there any cases with burns?" He said: "No. But the situation isn't very clear. My lads are feeling a bit nauseous."

'In fact, the fire was no longer visible, it had somehow crawled down the chimney. The covering, the roof, had fallen in . . .

'We talked with Kibenok near the power block itself, there where the firemen were standing. Pravyk, Kibenok, by then they had drawn up in two engines. Pravyk had jumped out, but he didn't come to me; Kibenok was rather excited, nervous.

'Sasha Skachok had already taken Shashenok from the power station. The lads had dragged him out. A beam had fallen on him, and he was burned. He died in the resuscitator on the morning of 26 April.

'We had no dosimeters. We were told there were gas

masks and protective suits, but there wasn't anything of the sort, it didn't work...

'I needed to make a telephone call. Kibenok said that he too needed to call his bosses; I went to the administration and social building, about eighty metres from the block. And I told the lads: "If you need any help, I'm staying around."

'I had had a real sense of alarm when I saw Kibenok and then, by the administration building, the operation lads. They were jumping from the third block and running to the administration building. You couldn't get any sense out of anyone.

'The door of the health centre was blocked...

'I phoned the central control panel. I asked: "What is the situation?" "The situation is unclear, stay around, give assistance if it is necessary." Later I phoned the medical unit. The deputy chief, Volodymyr Oleksandrovych Pecherytsia, was already there.

'I told Pecherytsia that I had seen the fire and that I had seen the fourth power block's collapsed roof. This was somewhere around 2am. I told him I was getting worried. I had come here but had so far done no work, and all the while the town depended on me. There might be urgent calls. I also told Pecherytsia that there were still no victims, but the firemen said that they were feeling nauseous. I began to remember my wartime hygiene, my time at the institute. Some knowledge filtered out, although it seemed that I had forgotten everything. What was the point of it for us? Who needed radiation hygiene? Hiroshima, Nagasaki, all that was so remote from us.

'Pecherytsia said: "Stay around there for the moment, ring again in 15-20 minutes, we'll tell you what to do. Don't get worried, here in the town we'll call out our own doctor. At this very moment three people came together to me. I think they had been dispatched with a young man of around eighteen. He was complaining of nausea, an acute headache, and was beginning to vomit. They had been working on the third block and, it seems, had gone over to the fourth... I asked him what he had been eating, when, how he had spent the evening, did he have a tendency to feel nauseous? I took his pressure, it was around 140 to 150 over 90, a little high and unstable, and the lad wasn't himself, he felt peculiar... I took him to the ambulance. There was nothing in the vestibule, nothing even to sit on, just two machines dispensing fizzy water, and the health centre was closed. I took him into the ambulance. And he "floated away" before my eyes, although he was awake, and at the same time such symptoms: confused, unable to speak. beginning to stumble, as if he'd downed a good dose of spirit, but no smell, nothing... Pale. And those who had been running out of the block had just kept clamouring "It's horrible, horrible," their minds were affected too. Then the lads said that that the apparatus needles were jumping off the scale. But that was later.

'I gave the lad injections of relanium, aminazine, and something else, and immediately I had done this three more arrived. Three or four from the operations section. It was all according to the textbook: headache with the same symptoms—a tight throat, dryness, nausea and vomiting. I gave them a relanium injection. I was alone, there was no medical assistant, and I immediately packed them into the car and sent them off to Prypiat with Anatolii Gumarov.

'I phoned Pecherytsia again, said how it was, what the symptoms were.

"Didn't he say that he would be sending you help straight away?"

"No, he didn't... I had just sent those off, when the lads brought me some firemen. Wearing capes. Several of them. They simply couldn't remain standing. I applied a purely "symptomatic" treatment: relanium, aminazine, to 'reduce' their mental problem and the pain... "

'When Tolia Gumarov returned from the medical unit he brought me a pile of drugs. I phoned back and said that I would not use them. After all, there were no burns cases. For some reason they loaded me with drugs. Later, when in the morning I came to the medical unit, no one wanted to take them off me. They had already started to measure me, and the background radiation was really high. I tried to give the drugs back, but they wouldn't take them. So I got them out, put them down and said: "Do what you want with them."

'Having sent the firemen off, I asked them to send potassium iodide, tablets, although there probably was iodine in the power station health centre. At first Pecherytsia asked: "Why, what for?," and then, apparently, when they saw the victims, they stopped asking questions. They got the potassium iodide and sent it. I began giving it to people.

'The building was open, but people went out. They were vomiting and feeling uncomfortable. They felt ashamed. I chased them all back into the building, but they just came out again. I explained to them that they had to get into the ambulance and go to the medical unit for observation. And they said: "But I must have smoked a little too much, I've just got too excited, there's been this explosion, and... " And they ran away from me. People just didn't fully realize what had happened.

'Later, in Moscow, in Clinic No. 6, I lay in a ward with a dosimeter operator. He told me that immediately after the explosion the needles on the power station apparatus had gone straight over the top. They had phoned the chief engineer and the security engineer, and the latter had replied: "What's all the panic? Where is the duty chief? When he arrives, tell him to phone me. But don't panic. The report is not in order," he answered and replaced the receiver. He was in Prypiat, at home. And later they jumped out with these "depeshky" (dosimeters—Iu.S.), and you couldn't get to the fourth block with them.

'My three ambulances were circulating all the time. There were already numerous fire engines; ours started to use their headlamps and sound their horns so they cleared the way for us.

'I didn't take Pravyk and Kibenok out. I remember, Petro Khmel was there, a dark-haired lad. I was in hospital with Petro for a while in Prypiat, our beds side by side, and then in Moscow.

'At six o'clock I too began to feel my throat hurting and my head aching. I understood. I was afraid. But when people see that close by there's a man in a white smock, they calm down. I had been standing there like everyone else, without a gas mask, without any protective gear.'

'But why without a gas mask?'

'Where would I have got one? I'd have rushed off to get one, but there was nothing anywhere. I phoned the medical unit: "Do we have any gauze?" "No." But that's enough. Working in a gauze mask? It does nothing at all. In this situation it was simply impossible to abandon people.

'At the block, when dawn broke, there was no longer any blaze. There was black smoke and black soot. The reactor was spitting, not all the time, but just like this: smoke, smoke, then an explosion. Fallout. It smoked, but there were no flames.

'The firemen had come down by then, and one fellow said: "Let it burn with blue flames, we won't go up there again." It was already clear to everyone that all was not well with the reactor, although the control panel simply did not give any concrete data. Just after five o'clock a dosimeter operator came in a fire engine. I don't remember who and where he was from. He came with firemen, they carried axes and broke down some door in the administration and social building, taking something away in boxes. I don't know whether it was protective clothing or equipment, but they loaded it into the fire engine. The dosimeter operator had a fixed appliance.

'He said: "How come you are standing there without protection? The level is very high here, what are you doing?" I said: "I'm working here."

'I came out of the administration and social building, and my ambulances were no longer there. I asked that dosimeter operator: "Where did that cloud go? To the town?" "No," he said, "toward Ianiv, it just barely touched our area in passing." He was about fifty, he drove off in a fire engine. And I felt ill.

'Then Tolia Humariv came again, something I'm grateful to him for. By then I was already making my way out, thinking I would ask the firemen for a lift so long as I could still move. My initial euphoria had passed, and weakness in my legs had taken its place. So long as I had been working, I had not noticed this, but now it began. A state of collapse, pressing on me, bursting, I was oppressed, with only one thought: to crawl somewhere into a corner. I thought neither of my family nor of anything, I somehow wanted to be left alone, just that. To get away from everything.

'Tolia Humariv and I stood there for five to seven minutes more, waiting to see if anyone would ask for help, but no one turned to us. I told the firemen that I was going to the centre, to the medical unit. If they needed me they should call me. There were more than ten fire engines there.

'When I came to the medical unit, there were a lot of people there. The lads had brought a bottle of spirits; drink, they said, you have to, it helps. But I couldn't, everything upset me. I asked the lads to take potassium iodide to my family in the hostel. But some were drunk, and others were running around constantly washing themselves. So I took a Moskvich—the driver wasn't ours—and went home. First, though, I washed and changed. I took potassium iodide to my family in the hostel. I told them to close the windows and not to let the children out; I told them all I could. I distributed tablets to my neighbours. And Diakonov, our doctor, came and took me away. I was given treatment at once, and placed on a drip. I began to "float away." I began to feel worse, and I remember everything rather hazily. Later I remembered nothing at all . . . '

That summer I received a letter from Donetsk from an old friend of mine, the dean of the Pediatric Faculty of the M. Gorky Donetsk Medical Institute, Volodymyr Vasyliovych Hazhiiev. Once, in the fifties, Hazhiiev and I had produced the Kiev Medical Institute satirical magazine *A Crocodile in a Smock*. It was popular among the students and teachers, and we drew caricatures and wrote witty captions . . . In his letter V.V. Hazhiiev told me about Valentyn Bilokin, a graduate of the Pediatric Faculty: 'During his years of study in the institute he was on the whole an average, ordinary student . . . He never tried to produce a favourable impression on his surroundings—teachers, administration and the like. He did the things he was charged with modestly, worthily and well.

'You felt he was reliable. In his studies he overcame his difficulties on his own, with no crises. He went worthily toward his set aim (he wanted to be a consultant pediatrician), doing all that was necessary. His natural decency, his kindness evoked a solid and deep respect particularly on the part of his group and course colleagues, and also his teachers. When in June we learned of his worthy conduct on 26 April in Chernobyl, the first thing we said was that he, Valyk, could have done nothing else. He is a genuine person, reliable and decent, a person to whom people reach out.

I met Valentyn Bilokin in Kiev in the autumn, when several things were already behind him: the hospital, a stay in a sanatorium, all sorts of fussing about getting an apartment and arranging work in Donetsk, every type of problem (the effort he had to make to receive the salary due him... for April, not to mention his getting the material compensation due to every inhabitant evacuated from Prypiat!).

Before me sat a lean, broad-shouldered, modest fellow, in whose every word and movement were restraint and a profound sense of dignity, both as a doctor and as a person. Only on the third day I learned by chance that he suffered from shortness of breath, although before the accident he went in for sport—weight-lifting—and could lift considerable weights. He and I went to Professor L.P. Kindzelsky for a consultation...

Valentyn told me about his children (he is the father of two little girls, the five-year-old Tetianka and tiny Katia, who was one-and-a-half months old at the time of the accident), and showed delight that at last he would work in the speciality which he had consciously chosen for himself in life: as a consultant pediatrician. And I thought how, that night, he had been the first doctor in the world to work on the scene of an accident of such a great scale, he had saved casualties gripped by fear, people stricken by radiation, how he had given them hope, because that night those had been his only medicines, stronger than relanium, aminazine and all the drugs in the world.

7

The Extraordinary Convoy

Aleksandr Iuriievich Esaulov, 34 years old, deputy president of the Prypiat Town Executive Committee:

'They got me out on the night of 26 April, just after 3 o'clock. Mariia Hryhorivna, our secretary, phoned me and said: "There's been an accident at the nuclear power station." An acquaintance of hers worked at the station and he'd called during the night, woken her up and told her.

'At 3.50am I was in the Executive Committee. The president had already been told, and he had left for the nuclear power station. I immediately phoned the head of staff of our civil defence, getting him up as it was an emergency. He lived in a hostel. He rushed over at once. Then the president of the Town Executive Committee drove up. Volodymyr Pavlovych Voloshko. We all got together and started to work out what to do.

'Of course, we had no idea what to do. As you say, it's only when the thunder bangs over your head.

'In the Town Executive Committee I'm president of the planning commission, I look after transport, medicine, communications, roads, works, building materials and pensioners. On the whole I'm young as deputy presidents go, I was chosen as recently as 18 November 1985. On my birthday. I lived in a two-room apartment. At the time of the accident my wife and two children weren't in Prypiat—she had gone off to her parents, because she was on post-natal leave. My son was born in November 1985. My daughter is six.

'So, I went to our Motor Transport Regiment and decided to organize the cleaning of the town. I phoned Kononykhin in the Executive Committee and asked him to send a street-cleaning truck. It came. What a situation! You won't believe this, but for our entire town there were four street-cleaning trucks! For fifty thousand inhabitants! In spite of the fact that the Executive and Town Committees were both on the whole aggressive, and had been to the Ministry and asked for trucks. Not because they foresaw the accident, but simply to keep the town clean.

'A truck with a tank drove up, where they'd dug it up, I don't know. The driver wasn't the usual one and didn't know how to work the pump. The water just flowed haphazardly from the hose. I packed him off, and he came back twenty minutes later, having learned how to work the pump. We started to clean the town by the filling station. By hindsight I now understand that this was one of the first procedures to follow to get the dust to settle. The water flowed in a soap solution. Later it turned out that this was a very polluted part of the town.

'At ten o'clock there was a meeting in the Town Committee building, a very short one, about 10-20 minutes. People weren't ready for a long talk. After the meeting I went straight over to the medical unit.

'I sat there in the medical unit. Suddenly I remembered: the block is clearly visible. It's right there, really close by. Three kilometres away. Smoke was coming from the block. Not exactly black... A sort of plume of smoke. Like a bonfire that had gone out, but from a bonfire like that the smoke is grey, and this was so dark. Well, and later the graphite started to burn. This was already nearer evening, there was a glow in the sky, just what we didn't need. There was so much graphite there... It was no joke.

'In the afternoon the second secretary of the Kiev Regional

Committee V. Malomuzh called me in and gave me the task of organizing the evacuation of the most seriously ill people to Kiev, to the airport, to send them to Moscow.

'From the headquarters of the national civil defence came Hero of the Soviet Union Major-General Ivanov. He flew in by plane.

'He handed over the plane for the transportation of people.

'All this took place sometime after 5pm on Saturday, 26 April.

'It turned out to be not so simple to set up a convoy. It's no easy matter loading up people. We had to prepare documents for everyone: medical histories, results of analyses. The main delay was precisely in processing these personal details. There were even moments like when a seal was needed, but the seal was at the power station. We crossed this bit out, and sent the person off without a seal.

'We took twenty-six people—that was one bus, a red interurban Icarus. But I had told them to give us two buses. Anything can happen. Heaven help us if there was some sort of delay and... And we needed an ambulance because there were two gravely ill people, stretcher cases, with thirty per cent burns.

'I asked them not to go through the centre of Kiev. These lads in the bus, they were all wearing pyjamas. Something of an extraordinary spectacle, of course. But for some reason they went along the Khreshchatyk, then left along Petrivska Avenue, and rushed off to Boryspil. They arrived. The gates were closed. It was night, around three o'clock, or just after. We sounded the horn and, finally, a spectacle worthy of the gods. Someone in slippers, riding breeches, and without a belt came out and opened the gates. We went straight out onto the airfield, to the plane. There the crew was already warming up the engines.

'And there was another episode that went straight to my heart. The pilot came up to me. And he said: "How much a dose have these lads had?" I asked: "What do you mean?" "Roentgens." I said: "Enough. But so what?" And he said to me: "Well, I want to live too, I don't want any excess roentgens, I've got a wife and children."

'Can you imagine?

'Off they flew. I said farewell and wished them a speedy recovery.

'We rushed back to Prypiat. It was the second day I hadn't slept, but I just couldn't. During the night, when we had been driving to Boryspil, I saw the convoys of buses going directly to Prypiat. To meet us. They were already preparing the evacuation of the town.

'It was the morning of 27 April, a Sunday.

'We arrived, I got some breakfast and went to see Malomuzh. I reported. He said: "Everyone who's been hospitalized has to be evacuated." The first time I'd got the most seriously ill out, now I had to get them all out. During the time I had been away, more people had turned up. Malomuzh told me to be at Boryspil at 12 o'clock. And our conversation took place around ten in the morning. It was simply unrealistic. All those people had to be got ready, all their documents had to be processed. And the first time I had taken twenty-six people, now I had to get 106 out, quickly.

'We got all this "delegation" together, processed them all and drove off at 12 o'clock. There were three buses, with a fourth in reserve. Icaruses. There were women standing there, saying goodbye, weeping, and the lads, all of them walking around, wearing pyjamas, and I was begging them: "Lads, don't get separated, I don't want to have to come looking for you." One bus was filled, then the second and the third, they were all in, and I ran off to the accompanying car, got in and waited five minutes, ten, fifteen; where was the third bus?

'It turned out that three more casualties had arrived, then more . . .

'Finally they arrived. There was a halt in Zalissia. We had agreed that if anything happened we would flash our headlamps. Going through Zalissia, there they flashed! The driver braked harshly. The buses stopped. The last bus was about eighty or ninety metres from the first. It stopped. A nurse flew out and ran with all her might to the first bus. It turned out that all the buses had their medics, but the medicines were in the first bus. She ran up: "There's a patient in a bad way!" And I saw Bilokin for the only time. True, at the time I didn't know his surname. They told me later that it was Bilokin. He was in pyjamas too, and he ran up with his bag to give assistance.'

V. Bilokin: 'The first group of casualties set off on the 26th in the evening, around 11pm and made straight for Kiev. They took the operators, Pravyk, Kibenok, Teliatnikov. We stayed behind for the night. On the 27th in the morning my doctor said: "Don't worry, you'll fly to Moscow. There's been an order to get you out before lunch." When they were taking us out in the buses, I felt bad. We even stopped somewhere just after Chernobyl, someone was getting worse, and I ran out and tried to help the nurse.'

A. Esaulov: 'Bilokin ran, and they tried to catch him. "Where are you going, you're ill." He was a casualty too... Off he rushed with his bag. And the most interesting thing was that when they started to rummage in that sack, they just couldn't find any ammonium hydroxide. So I asked the traffic officials who were accompanying us: "Have you any ammonium hydroxide in your first-aid kit?" "Yes." We swung round and drove up to the bus. Bilokin immediately put an ampoule under the lad's nose. The trouble eased.

'And another incident in Zalissia comes to mind. The sick people had got out of the buses, some for a smoke, others to stretch themselves walking here and there, and suddenly a woman ran up shouting, creating a real uproar. Her son was in this bus. That was all we needed! What a coincidence! Can you imagine? Where she had appeared from I simply could not understand. He called her, "Mother! Mother!" and calmed her.

'At Boryspil airport a plane was already waiting for us. Polivanov, the airport chief, was there. We drove out onto the field to get up close to the plane—the lads were all in pyjamas, it was April and not hot. We drove through the gates, onto the field and a little yellow Riga truck rushed out, its driver swearing that we had gone out without permission. We had first of all gone to the wrong plane, anyway. The Riga truck accompanied us.

'And there was another incident. I was sitting calmly with Polivanov, there was a pile of telephones, we were processing documents for the transportation of the casualties. I gave them a Chernobyl nuclear power station receipt and a letter of guarantee that the station would pay for the flight, it was a TU-154. A good-looking woman came up and offered us coffee. Her eyes were like Christ's, and she clearly knew what was going on. She looked at me as if I was someone come out of Dante's inferno. It was already the second day I hadn't slept and I was terribly exhausted... She brought the coffee. Such a little cup. I drank this mouthful in one go. She brought me a second. Wonderful coffee. We got everything done, I stood up, and she said: "That'll be fifty-six kopecks." I looked at her, I just couldn't understand. She said: "I'm sorry, but here such things are done for money." I was so far away from thinking about money, from all that... It was as if I'd come from another world.

'We washed the buses again, took a shower, and off we went back to Prypiat. We drove out of Boryspil around four o'clock in the afternoon. Along the road we met buses...

'They were bringing the people of Prypiat out.

'We arrived in Prypiat. It was already an empty town.

'That was on 27 April 1986, a Sunday.'

8

Before the Evacuation

L. **Kovalevska**: 'I lived in the third microraion. I often had insomnia and took things to help me sleep. On 25 April, Friday, I had just finished my narrative poem "Paganini." I had been working on it every night for three months. And that night I decided to take a rest. So I took something to help me sleep. And I slept the sleep of the dead. I didn't hear the explosion. And yet where we live, if an exhaust backfires, you hear it. Even the windows rattle. In the morning my mother said: "Either there was something that made a noise like thunder at the power station last night, or jets were flying all night." I didn't ascribe any significance to this. It was Saturday. I began getting ready to go to our literary club. I once ran it, it's called "Prometheus." Power workers from the station and construction workers went to it. . . I went out and looked—all the roads were flooded with water and there was some white solution, everything was white, foaming, all the roadsides. And I know how it is when there is an accident with an accidental spill. I began somehow to feel uneasy. I walked on. I looked: policemen here and there, never had I seen so many policemen in the town before.. They were not doing anything, just standing around near strategic buildings: the post office, the Palace of Culture. It was like a war situation. It suddenly hit me. People were out walking, there were children everywhere, it was so hot. People were going to the beach, to their dachas, fishing, many people were already at their dachas, sitting by the river, near the cooling pond (that was a sort of artificial reservoir near the nuclear power station)... I didn't get to the literary club. Iana, my daughter, had gone to school. I returned home and said: "Mother, I don't know what has happened, but don't let Natalochka (that's my niece) out, and when Iana gets home from school, make her stay at home immediately." But I didn't tell her to close the fortochka. I didn't think of that. And it was so hot. I said: "And don't you go out, and don't let the girls out today." "But what's happened?" she said. "I don't know anything exactly, I just have a feeling." I made off there again, to the central square. Our literature students were there, and I went to meet them. You could see the reactor well, you could see it was burning and that a wall had split open. There were flames rising over the hole. the chimney between the third and fourth blocks was red hot, like a fiery pillar. Flames can't stand like that, without swinging to and fro, but there stood a fiery pillar. Whether or not it was flames coming from the opening, I don't know.

'The whole day we knew nothing, and no one said anything. Well, it was a fire. But as for the radiation, that there were radioactive emissions, nothing was said about that.

'And Iana came from school and said: "Mom, we had physical training for almost an hour in the street." Madness... '

Aneliia Romanivna Perkovska, secretary of the Prypiat Komsomol Town Committee:

'Already in the morning there were fellows coming into the Town Committee with offers of help. But we didn't know what to do ourselves. There was no information at all. There were just rumours...

'They had put the fire out. But what about the reactor?

'They even argued about whether the reactor had been exposed or not. Nobody could believe that the reactor had been exposed. We chatted with people who had studied these reactors. Surely the reactor in itself was so well conceived that even if you had wanted to blow it up, you couldn't have. It was difficult to believe that it had been exposed.'

Iurii Vitaliiovych Dobrenko, 27 years old, instructor at the Prypiat Komsomol Town Committee:

'Next door to me in the "Jupiter" works hostel lived a doctor, Valentyn Bilokin. He worked on the ambulance service. I used to go fishing with him, he was a fine fellow, we both went in for spinning. When that accident happened, he was on duty, and later he came to the hostel, undressed, just wearing his white doctor's smock; he came somewhere around six in the morning, he had been in the neighbouring block, distributing iodine and pills. He said: "Take them just in case." Then the ambulance came for him, and they took him away. They didn't wake me up. They told me all this in the morning.

'There was a sort of vagueness in our Town Committee the whole day. But after six in the evening we all got together again—we now had a concrete job to do... '

A. Perkovska: 'About four in the afternoon on Saturday, 26 April, the members of the Government Commission began to gather. The idea had been proposed of loading sand onto helicopters and scattering it over the reactor. Whose idea it was, I can't say. The debates there lasted a long time. Was lead necessary? Was sand necessary? Commands were very quickly given and then countermanded. It was obvious why, there had never been a situation like it before. Something fundamentally new had to be found.

'Finally it was decided to load sand. We have a cafe, the "Prypiat," near the river station. They washed sand there for the sixth and seventh microraions. Really good-quality sand, clean, unadulterated. The sand was loaded into sacks. There were many people who had been posted to the town. Lads ran up from Ivano-Frankivsk. And they said: "We need a propagandist!" That sounded really military. "We need a propagandist, the guys there are already exhausted." The guys there had already worked very hard...

'Rope was needed to tie the sacks. There was none left. I remember we took some red material, red calico, which was lying around for festivals, and we began to tear it up into strips.'

Iu. Dobrenko: 'So in the evening we all met in the Town Committee, and I was given the first task: spades were to be brought to the "Prypiat" cafe; we had to go and get those spades, around 150 of them, and the others went to the hostels to collect young people. The young people came. Around eleven in the evening a truck came with empty sacks, and we started to fill the sacks with sand.

'One of the first to come was Serhii Umansky, the secretary of the Komsomol organization of the Prypiat assembly-plant administration. He worked without a dosimeter, without anything; I now remember they gave him a white suit, and he worked through the night; later they let him off to get some sleep, and in the morning I saw him again in the Komsomol Town Committee: "What's this, is it you, Serhii?" "Yes," he said, "I'm working, we're filling sacks".'

Iu. Badaev: 'We spent the whole night shift of the 26th at the power station. Around eight o'clock there came the command: everyone to leave their places of work. We went in the civil defence premises. Later they let us go home.

'I told my wife that something really bad had happened, from our window you could see the ruined block. I said: "It's advisable not to let the children out anywhere. And the windows should be closed." My wife, unfortunately, didn't do as I asked, she felt sorry that I had endured such a lot. I went to bed and she let the children out, so there'd be less noise. She gave me a chance to rest... It would have been better if I hadn't slept. On Saturday there was no instruction not to let children out, That came on Sunday. At around ten o'clock a woman came and told us not to let the children out, not to go out of the house, and to listen to the radio. At two o'clock the evacuation began.

'And a few months after that, very early in the morning, while it was still dark, when infernal mist hung over all the world, Sashko Esaulov and I drove out into the Zone. We went in a Zhiguli-008. Esaulov is an excellent driver, and our incursion was like being on a rally. The headlamps full on pushed obstinately into the impenetrable clouds of mist and blinded us! Dipped, almost nothing was visible, especially where there were no road markings, and from time to time the road went away from us, and then somehow came back. The car handled really well. ("That's what comes of having a front-wheel drive!" Sashko, a dark-eyed and ruddy sturdy lad, said ecstatically). In my lap I had a cassette recorder, and all along that tense and dangerous road, buoying us up and giving us confidence, the lads from Liverpool, the famous "Beatles," sang in the car. Their music blended somewhat strangely with the difficult journey through the Zone. There were no normal number plates on our car; just on the hood and the sides you could see the large numbers: 002, like on racing cars. Occasionally we came across armoured troop carriers on the road, their headlamps on, and here and there chemical defence sub-detachments were working: soldiers in black suits and special vivid green slippers.

'Near the entrance to Prypiat we suddenly saw, on both sides of the road, piles of sand cut through by bulldozers, uprooted stumps, and further on a rust-coloured, as it were burnt, pine wood. This is the sadly famous wood which has already entered the legend of the Zone under the name of "Red Forest." It was here that a part of the debris from the fourth reactor was deposited. Behind this wood the town started, and by its side stretched the area of the so-called "Nakhabivka," dachas put up without any planning, miserable wooden hovels, with meagre little gardens. That Sunday people were taking a rest there.'

A. Perkovska: 'On Saturday in school No. 3 an assembly of the school team was fixed; it's a big school, 2,500 pupils, so their team has 1,500 children. The meeting was to be held in the Palace of Culture. Well, at the meeting on the morning of the 26th they put this question to this assembly, and the Second Secretary of the Regional Committee of the Party, V. Malomuzh, told us to carry on as we had planned. When we left after this meeting, the headteacher said to me: "Well, what am I to do?" I said: "Hold it in the school.

There's no need for all the children to be there." And the assembly did take place, only in the school's sports hall.

'On Saturday all the lessons took place, nothing was cancelled. But there were no outside activities, races and the like.

'In schools No. 1 and 2, wherever I went the windows were closed. There were wet cloths on the floor, people on duty by the doors, and no one was let in or out.

'I don't know about other schools.

'A run, "Health," had been planned for Sunday. The teachers didn't know whether it would take place or not. One teacher phoned the Town Committee: "I'm assembling all the children in the school in the morning." And when they told her that everyone was already shouting about an evacuation, she blurted out: "What evacuation, you lot? But today we've got our 'Health' run!.."

'Just imagine: ninety minutes remained to go to the evacuation. Our children's cafe, in the big shopping centre, is brimming with parents and their children, who are eating ice-cream. It's a day off, everything's fine, everything's calm. People are walking about the town with their dogs. And when we go up to people and explain, the reaction is stormy and suspicious: "What business is it of yours that I'm taking a walk? If I want to take a walk, then I will do." And that was that. That's how people understood it.

'I remember how Sasha Serhiienko, our second secretary, flew into a rage: "What have I just seen? A child sitting in the sand, and the father is a senior power engineer. How could he, knowing there has been an accident at the nuclear power station, let his child sit scratching around in the sand? And the street is right next to that wood." He had "Red Forest" in mind.'

9

The Evacuation

Forty-five years after the beginning of the Great Fatherland War that terrible word 'evacuation' rang out again in our country.

I can remember Kiev in forty-one, possessed by anxiety, the hubbub at the rail station—we lived near the station. Some people were leaving, others were staying, still others couldn't believe that the Germans would come to Kiev (on the first day of the war my father said that in two weeks we would be in Berlin), and even others had already laid in provisions, getting ready for the occupation. No one really knew, and so the uncertainty just increased. And German planes were flying over Kiev, and under the plexiglass bubbles you could see the heads of the radio operators-gunners, victoriously looking down on the ancient capital of Rus', on its cathedrals' cupolas resplendent with gold. Ominous rumours went around the town: that we were surrounded, that there were saboteurs, that there had been parachute drops and that German tanks had broken through; and we, doomed, sat there at home, going nowhere, because father had gone to the front zone and we had heard nothing of him. My father was a roads engineer: he worked in the highways administration of the NKVD, he was a member of the Party, and we could only guess what awaited the family of a member of the NKVD and a Bolshevik during an occupation. But on one anxious scorching day in July 1941 a one-and-a-half ton truck drove up to our building on Solomianka, father jumped down from it and gave us thirty minutes to get ready. Mother ran about the apartment, not knowing what to take. My father said that it wouldn't be for long, just a month, two at the most, until autumn, and so we didn't take any warm things, and in the rushing about we forgot the most indispensable things. Later, in the severe Russian cold of Saratov we recalled father's optimism, an optimism educated by newspapers, radio programmes and films: before the war there had been a marvellous film: 'If There is War Tomorrow.'

From that time I have perceived an evacuation—whatever its scale—as a great calamity, always unexpected, which always causes shock and perplexity irrespective of whether it has been well or badly organized. Some sort of historical hurricane tears people from their roots in their native soil, and it is very difficult to start life anew along its accustomed tracks.

A. Perkovska: 'The first talk about the possibility of an evacuation came on Saturday evening, around eleven o'clock. And at one o'clock in the morning we had already been given the task: in two hours the documents for transfer out were to have been processed, They left me, telling me to get the documents ready to be handed over. It was like a whiplash across my nerves, just like the war.

'Every one of us in the Town Committee, has kept till now this sense of a boundary: before the war and after the war. We simply say: that was before the war, We know very well that this was before 26 April and that was after, if we need mentally to locate something.

'I began to think: what did we need to take out? Clearly: banners, seals, stock cards. But what else? In our instructions the word "evacuation" just does not figure. Nothing had been foreseen for such an incident. But we had three other committees subordinated to the Regional Committee. What would happen to their documentation? The Komsomol construction committee of the nuclear power station is situated opposite the fourth block in the administration building. And the power station committee is a little further away. You couldn't get there.

'I summoned the accounts section deputy, and a statistician. This happened at night. They came quickly and we began to wonder what to do.

'We got all the documentation together. There was no time to make any calculations. We put it all in sacks and sealed them. In the section Svitlana and Masha worked, and some young men helped them. And we helped in the Town Committee of the Party. The Government Commission was working there, and they had certain questions which they couldn't solve, and they needed a local person. Whom should we summon, whom should we telephone? Later we went to the Executive Committee and I was given the plan for the fifth microraion. I had to carry out the evacuation there.'

L. Kovalevska: 'During the night of Saturday to Sunday we had gone to bed when the doorbell suddenly rang. It was about three o'clock. The neighbours' girl: "Auntie Liuba, wake everyone up, get ready, there's going to be an evacuation." I switched the light on and listened; people in the entrance to the building were weeping, bustling about, our neighbours were getting up. My mother got dressed. She was shaking. I said: "You see, I turned on the radio, and it's silent. If there was anything going on, they would announce it on the radio." "No, I'll wait a little." She waited for thirty minutes. Nothing. An hour. Still nothing.... "Mama," I said, "go and lie down. You can see there's nothing happening." Then people got up, there were tears, someone had gone off during the night to Chernihiv, by the Chernihiv train at four o'clock in the morning, from the Ianiv station.'

A. Perkovska: 'At 5am on 27 April they let us go to pack our things. I arrived home. My brother was sitting in an armchair, he wasn't asleep. I told my brother to pack something at least. Well, he packed his papers. You see, I'd known about the evacuation earlier and let my brother know. But look, he just took his papers and a spare shirt and jacket. That's all.

'And I did the same. I left with such a little bag. I took my papers, and that was all. And later it turned out... When the dosimeter operators checked our clothes, they had to be changed, and it turned out that we had nothing to wear. Girls in the Ivankiv regional committe gave me some clothes. Overall, in the rush people took nothing with them. All the more as people had got used to "believing." They told us it would be for three days. Although they knew full well it would not be for three days, but for longer.

'I think it was quite proper that they said what they said. Otherwise the evacuation would not have been carried out so quickly.

'When I had arrived home in the morning, the neighbours started to call after half an hour. I calmed them as best I could. I didn't say there wouldn't be an evacuation. And I didn't start to say that there would. I said: "Get ready and wait for news." I managed to drink some coffee and around six in the morning went back to work. There the word "evacuation" resounded in a more real way. There was consultation among us on the text to notify the people of Prypiat. I can more or less recite it from memory:

'"Comrades, in connection with the accident at the Chernobyl Nuclear Power Station we announce the evacuation of the town. Have your papers, indispensable things and, if possible, rations for three days, with you. The evacuation will begin at 1400 hours."

'It was broadcast four times.'

L. Kovalevska: 'I said to my mother: "If there's going to be an evacuation, it means it's not for three days. there's no such thing." I took all the warm things for the children. Two bags, products. And the refrigerators were crammed full, so much money gone to waste, my mother's pension, my wages. After all, it was nearly May Day, and then 9 May. I scraped together all I could: threw some things into the garbage, all the products, switched off the refrigerator, covered everything, put what I had cooked for the children into bags. We still had 150 roubles, I took them. I took a warm shawl for mother, for myself a jacket, pants, nothing else.

'While mother was sitting weeping, I said: "Wait, don't bother me, I'll get all the papers together." And I took my poetry. The rough work, in notebooks, I took it all and packed it. I had *The Diplomat* among my things.

'"And now," I said, "I need nothing".'

A. Perkovska: 'We rushed into the fifth microraion to see the evacuation through. There was Oleksandr Fedorovych Marynych from the Town Committee of the Party, and myself from the Komsomol Town Committee.

'Here I'd like to single out one interesting girl. Maryna Berezina, a student of the biological faculty, worked with me as a senior group leader. Her husband had been working in the fourth block precisely on that shift. On Saturday she had no idea where her husband was and what had happened to him. His surname was Berezin too. And on Sunday I met her. She ran up, and I said: "Well, Maryna, have you any news of your husband?" She said: "I know he's alive, but I don't know any details."

'And she said: "Should I help you?" She didn't live in my microraion. I told what there was to do, and this girl passed the whole main evacuation with us, she didn't even leave herself. She said: "Nelia Romanivna, if you need help, I'll be at home and will come to you in the Town Committee."

'Later, when we carried out the main evacuation, they began to find and take away the people who were hiding and didn't want to leave. Well, Maryna phoned me and said: "They've come for me! Can I go now or not? What a girl!"'

Iu. Dobrenko: 'I was responsible for the evacuation of a microraion. I co-ordinated the work of the police, housing and maintenance, and transport. What were we afraid of? Well, that there'd suddenly be a jam somewhere or panic. Let me tell you: the evacuation went through in a very organized way.

'People calmly came out with little bags, according to the radio instructions; they assembled near the entrances and buses quickly began to come up to each entrance; a policeman took a register, the people got into the bus, and off they went. In my district there were approximately 15,000 inhabitants, and we finished the evacuation in an hour and fifteen minutes. What problems were there? We tried to persuade, to beg people not to take prams, but they took the prams, because they had little children, no one would listen to us. No one tried to take any big and heavy things. On average there were two bags per person.

'The young people also conducted themselves in an organized way.

'What other problem did I have? Three hours before the evacuation a man died in my microraion. He was old and had been ill for a long time. It was a young family, with two children, and this grandad lived with them. But they had to be evacuated. Well, we solved the problem: he was taken to the mortuary. Some people had remained on duty at the medical unit and they helped, and buried him.'

A. Perkovska: 'We evacuated our microraion, No. 5, last. The people were in the street for such a long time . . . It was so scorching. And the radiation level was rising. So you ask: "Comrades, take your children into the entrances." They listened and took their children in. I went off, and from two houses away looked back—the children were back outside. They said to me: "It's stifling in the entrance. You try and stand in an entrance for hours on end".'

The evacuation.

That sunny Sunday, 27 April, thousands of Kiev people were getting ready to go to the countryside, some to their dacha, others to go fishing, and others to visit relatives or friends in suburban villages. But something, clearly, had gone wrong in the ordered transport system of the town, because some of the routes had been cancelled and only one or two buses were guaranteeing others. Crowds built up at the stops, people swore and rebuked the remiss controllers in the bus garages.

In Kiev that day hardly anyone knew of the calamity which had occurred 148 kilometres to the north. The majority of the people of Kiev did not know that on Saturday the buses had been put on alert and during the night convoys of buses had moved from Kiev and the Kiev area toward Prypiat. These were ordinary buses used on town or suburban routes, a lot of yellow Icaruses with a trailer and an 'accordeon.'

The evacuation

Imagine a convoy of a thousand buses with headlamps on, going two abreast along the highway and taking the many thousands of the population of Prypiat out of the afflicted zone: women, old folk, adults and newly-born babies, those sick from common ailments and those who had suffered from radiation.

Imagine the people who were leaving their clean, young, wonderful town, of which they were proud, in which they had already put down roots and brought children into life.

They were given a cruelly limited amount of time to get ready, they were leaving their homes (as became clear later, for ever) and they were leaving just as they were, dressed in summer clothes, holding on to the most indispensable things. But it turned out that the most indispensable things were more often left behind. A comprehension of what was most necessary, most important for a person came later. And later, when it became possible to return to their apartments and collect certain things, which were then subjected to the harsh dosimeter check, people, as it were more mature now, rushed not to take their 'prestigious' rugs (the pile in carpets took up a great deal of radiation), not to their crystal glassware, but to those things which presented a spiritual value: photographs of people close to them, favourite books, old letters, certain ridiculous but memorable trinkets, things which constituted the deeply personal and very fragile world of a person, who lives not only for the present, but also for the past and the future.

Everyone unanimously, both those evacuated and the doctors whom I met shortly after the evacuation, asserted that there had been no panic. People were silent, their minds were concentrated, sometimes they were in a state of shock and, as it were, a slowed-down state, still not understanding what had happened and because of that feeling strangely carefree. I met such people. There were hardly any tears or petty conflicts, no one tried to impose their rights. It was only in the eyes that you caught pain and anxiety.

The convoys of evacuated people moved to the west, to the villages of the Polissia and Ivankiv areas, which adjoin the Chernobyl area. The Chernobyl area itself was evacuated later, on 4 and 5 May.

The evacuation.

The mass exodus of thousands of people from their places of settlement created numerous complex problems—organizational, social, moral. Nothing was simple, and it is quite wrong to paint these events in rosy colours alone. Of course, the newspapers those days, describing the benevolence with which local people received those evacuated, were not trying to deceive anyone. It was like that, it was a fact. Ukrainian Polissia, whose inhabitants are called Polishchuky, manifested their eternal traditional national features: gentleness and kindness, hospitality and sensitivity, a desire to help people who had fallen into misfortune. But this is only half of the truth. Because it must be clear to everyone that a certain disorder and confusion reigned in the Polissia and Ivankiv areas at the beginning of May. Parents were looking for children, wives for husbands who had been working at the power station on the day of the accident, and from all the ends of the Soviet Union anxious telegrams from relatives and friends flew into the now non-existent post office in Prypiat...

I remember how during those days I went to the Ivankiv cultural centre, and again my heart ached, again the days of the war came back to my mind: in the rooms there lay mountains of overalls—white and grey—people were crowding around the notice board, queuing up in the information centre, asking each other about acquaintances, eagerly listening to the local radio announcements. Information was worth its weight in gold. Such a prosperous, calm and apparently immutable life had been torn loose from its anchor and now followed the current in an unknown direction... The same happened in Poliske too. The wall of the Regional

Committee of the Party became a new type of information office: here you could find the addresses of organizations evacuated from Prypiat, the addresses of acquaintances, and learn the latest news.

L. Kovalevska: 'Our bus didn't get to Poliske. We were found places in Maksymovychi. Later, when we arrived in Maksymovychi, the dosimeter operators took measurements: the radiation level had risen there. Let's get them out of there quickly. There echoed the watchword: first pregnant women and children. Just imagine the state of a woman who came to a dosimeter operator, and who measured the child's shoes: "Dirty." Pants: "Dirty." Hair: "Dirty"... When I had sent my mother to Siberia with the children, I felt better.

'And on 8 May, when I arrived in Kiev and Sergei Kiselev, the *Literaturnaia gazeta* correspondent in Ukraine invited me to spend the night at his place, I took a bath, I ran the water and I relieved myself by weeping. And I wept at the table. I felt so much pain for people, for the lack of truth. The newspapers were writing lies. Perhaps it was the first time I had come close up against this... To know the real essence of things and to read such bravura articles. It was a terrible shock and deeply upset me... '

A. Perkovska: 'After the evacuation I remained behind in Prypiat. That night, when everyone had already gone, I came out of the Town Committee. The town was in darkness. It was really dark, you understand. There were no lights anywhere, no lights in the windows... At every step there was the military police, checking documents. Once out of the Town Committee, I got a certificate, and went to the entrance to my building. I arrived; in the entrance there was no light either, I went in, into the dark night, and up to the fourth floor. I have a cosy apartment, but it was, as it were, no longer my own. That was a terrible shock.

'On Monday, the 28th, we went out to Varovychi, to have the party assembly. We spent the whole night there. Hardly had I arrived than we began to register people for Village Councils. So much was unclear. At last they assembled the communists and then the Komsomol members. The next day

I went to Poliske, returned to Varovychi, and later was taken to Ivankiv. There they were organizing a headquarters, and there were people I knew: from the Town Committee of the party Trianova, Antropov, Horbatenko, from the Executive Committee Esaulov, and from the Town Komsomol Committee, myself.

'I worked there from 8am to 9pm, both in the headquarters and going around the villages. There were crowds of people, some looking for their children, others for their grandchildren...

'The fact is that there was no evacuation scheme, and we did not know in which villages were which Prypiat buildings or microraions. Even now I don't understand: according to what scheme were people evacuated, who went where? In Poliske we had a list of children. So I would phone the Village Council and ask: "Do you have such and such parents? Their children are looking for them." And they could say to me: "We have such and such children who are without parents. Generally, we do not know where these children are from." You sit and phone all the Village Councils. Sometimes it would turn out that in some village a dear old grandmother was sitting with someone else's child and had said nothing to anyone...

'It was necessary to take to children off to Pioneer Camps, and later women with pre-school children and pregnant women. You had to note their numbers and where they were to be taken. We had our Komsomol assembly, appointed Komsomol organizers, so that at least there was a person on whom people could rely and with whom they could be in contact.

'It was different in those days. One man comes to mind. I would like that man to read these words of mine, to arouse his conscience. It was 1 May. I arrived at the information centre early. None of our people were there yet. A man of about forty-eight was standing there and he said: "Ah, so are you from the Prypiat Town Committee of the party?" "Yes, I am." "Give me a list of those who have died." I said: "Two people died: Shashenok and Khodemchuk." "That's a lie." I said: "What basis have you for talking to me like that?" And he shouted: "Of course, you're beautiful, you're in full bloom (I was standing there in someone else's clothes), you're so calm, because you've taken everything away from Prypiat. Do you think we don't know? We know everything!"

'At that moment there was only one thing I wanted: to plonk that man in a car, take him to my apartment, and give him a good talking to...

'His son had been working at the power station. So I said: "As far as I know, he is in the Pioneer Camp Kazkovy." And again he shouted: "How can you talk to me like that, I am a miner, a person of distinction." I asked him: "Where have you come from?" He replied: "Odessa."

'We gave him a car and he went to "Kazkovy," he found his son there, just as I had told him; then there were sincere thanks, but he still hadn't really understood. I remembered this for a long time. His conduct so upset me that I couldn't recover my wits for several hours.

'Of course, there was much discord and many difficulties, but I would say that our people, from Prypiat, behaved in a worthy fashion.'

The evacuation...

It is true that it was carried out in an organized and clear way. It is true that courage and steadfastness were displayed by most of those evacuated. All this was so. But are the lessons to be learned from the evacuation really limited to this? Surely we will not begin to gratify and calm ourselves with half-truths, closing our eyes to the home truths which were revealed during those days? Will we really succeed in quelling, in muffling the bitter questions of thousands of people through our being organized and disciplined? These are questions directed to those who had a duty to be dir]ected not by the cold and indifferent calculations of a cowardly bureaucrat, but by the ardent heart of a citizen, a patriot, a communist, who takes responsibility for the life and health of his people, for his future—the children.

After the publication of my one 'Chernobyl' article in *Literaturnaia gazeta* the editors sent me a letter. Here it is:

'This is a letter from workers from the town of Prypiat (we now live in Kiev). This letter is not a complaint, but just isolated facts from which we ask you to draw conclusions. We adduce examples of the criminal irresponsibility of officials from Prypiat and Kiev. First of all this irresponsibility was demonstrated as regards all the children (in the thirty-kilometre zone), when for a whole twenty-four hours before the evacuation nothing was announced, children were not forbidden to run around and play outside. We, knowing the level of radiation because of our work duties, telephoned the civil defence headquarters of the town and asked: "Why are there no instructions regarding the conduct of children in the street, regarding the absolute need for them to stay indoors and so on?"

'They replied: "This is not your affair... The decision will be taken by Moscow... " And only later (7 May 1986) did everyone learn that the decision to evacuate and dispatch their children to the Crimea (their grandchildren and their grandmothers) had been taken by the "Higher Leadership" without delay, and "selected" children had been sent to Crimean sanatoria on 1 May.

'Here is a second example of irresponsibility, when at a difficult moment it was necessary to make urgent use of indispensable property, apparatus to check the situation. The indispensable property turned out to be unusable. How does one evaluate this? Why did the leaders, occupying high positions and several years in a row receiving (unearned) wages, not know the real state of affairs—regarding the same property of the civil defence and with other outrages? Why did they not check, why were they satisfied with bits of paper about "complete prosperity"?

'We ask you to check everything with the State Commission and to take the necessary measures, particularly as regards those sore questions where dishonesty and official ineptitude on the part of the "great leaders" are at fault.

'Our address: Kiev, Main Post Office, general delivery. (The letter was written in June 1986, when the people evacuated from Prypiat had still no fixed address—**Iu.S.**).

'Signatures: Nikulnykov, S.V., Kolesnyk, D.V., Pavlenko, A.M., Radchuk, N.N.'

The authors of the letter touched, among others, on one of the sorest questions of the Chernobyl epic: the timeliness and quality of the measures taken to protect people from the effects of the accident. This question continues to worry many thousands of people, until now it resounds in credulous conversations within the narrow circle, within the family, but for some reason a shameful silence has been drawn over it in the public statements of leaders at the town, area and republican levels. It seems to me that the interests of glasnost—this very important factor in the restructuring of our society in the spirit of the decisions of the XXVII Congress of the Party—require a fundamental and open review of this problem. The time has come to remove the cloak of secrecy. If the authors of the letter and those who agree with them (and there are tens of thousands of them) are wrong about something, if everything was done perfectly, then this needs to be convincingly proved and explained. I fear, however, that it is difficult, perhaps even impossible, to do this.

I do not take upon myself the role of judge or accuser. Now, many months after the accident, it is easy to shake one's fist. I don't want to assume the pose of an omniscient public prosecutor. But I want somehow to understand: what exactly did happen? Many people from Prypiat (let us recall A. Perkovska's story) will never forget the meeting which was conducted during the morning of 26 April in Prypiat by the second secretary of the Kiev Oblast Committee of the Party, V. Malomuzh, who gave the instruction to do everything to continue the normal life of the town, as if nothing had happened: schoolchildren had to study, shops to be open, and the weddings of young people, scheduled for the evening, had to take place. To all the perplexed questions there was one answer: that's how it has to be.

For whom? And in whose name? Let's discuss this calmly. From whom did the calamity have to be hidden? By what legal or ethical considerations were the people guided who took these more than doubtful decisions? Were they aware of the real dimensions of the catastrophe? If they were, then how could they give such instructions? And if they weren't, then why did they hasten to take upon themselves such a serious responsibility? Surely, in the morning of 26 April, the radiation levels were known, levels which had risen steeply in consequence of the expulsion of fuel from the power station? I can remember how during those May days in one of the Kiev hospitals one could see a woman, an inhabitant of Prypiat, who on that fateful Saturday, like thousands of other townsfolk, had been working on her personal plot near 'Red Forest,' I have already recounted its story. She had radiation burns diagnosed on her legs. Who will explain to her in whose name she endured these torments?

And the schoolchildren who, knowing nothing, were let outside on Saturday during the breaks. Surely it was possible to look after them, to forbid them to be outside? Would anyone really have condemned the leaders for such 'over-insurance,' even if it had been superfluous? But these measures weren't superfluous, they were absolutely indispensable. There is an irony of fate: three days before the accident civil defence exercises were done in the Prypiat schools. They taught the children how to use the protective equipment for individuals: cotton-gauze masks, gas masks, how to decontaminate. On the day of the accident none of these measures, even the simplest, were applied.

On account of the situation of secrecy which reigned in Prypiat immediately after the accident, things got to the point where even the responsible workers in the Town Executive Committee and Town Komsomol Committee did not know the real levels of radiation for two whole days. They had to make do with the rumours which went round the town, with the vague hints of acquaintances, with the meaningful expressions of dosimeter operators... Are you surprised that in such a situation of total 'suppression' of information numerous people gave in to the rumours and rushed from the town along the road which led through 'Red Forest'? Witnesses relate that along that road, which was already 'glowing' from the full blast of radiation, women went pushing prams...

Perhaps, bearing in mind the extraordinary and unexpected nature of the situation, it was impossible to do otherwise? No. Specialists assure us that it was possible and necessary to do otherwise: it only took the local radio to announce the possible danger, to mobilize the town's emergency services to take containment measures, not to let out into the street those who were not working to eliminate the consequences of the accident, to close the windows, to apply immediate iodine prophylaxis to the population. Why was this not done?

Clearly, because the doctrine of universal prosperity and obligatory and immutable victories, joys and successes, which entered, over the last decades, the heart and soul of many leaders, here played a fateful role, stifling in them the voice of doubt and the orders of professional, party and civic duty: to save people, to do everything humanly possible to prevent the calamity.

In this situation an unattractive role was played by the then director of the nuclear power station, Briukhanov who, earlier than others and better than others, understood what had really happened at the station and around it. The extent of his guilt will be established by the organs of justice. But one cannot simply transfer to Briukhanov the sins of other officials.

After all, there is moral justice: how could it happen that the Prypiat doctors, the directors of the medical unit V. Leonenko and V. Pecherytsia, who were among the first to find out about the extremely unfavourable radiation situation (by the morning dozens of people with a serious form of radiation illness had been brought in to the hospital), how come they did not sound the alarm, shout out loud at the assembly that Saturday morning about the calamity that was drawing near? Surely the falsely understood consideration of subordination, of unconditional and unthinking execution of 'instructions from above,' the adherence to imperfect and pitiable official instruction did not stifle in their hearts their loyalty to the Hippocratic oath, the oath which for a doctor is the highest moral law? Moreover, this applies not only to these, on the whole ordinary doctors, but also to many more senior doctors: let us at least remember E. Vorobev, the former Deputy Minister of Health of the USSR. However it may be, today it is clear that the mechanism for taking responsible decisions, connected with the protection of people's health, has not withstood serious test. It is cumbersome, many-layered, too centralized, slow, bureaucratic and ineffective when events are developing swiftly. Innumerable agreements and co-ordinations led to almost twenty-four hours being spent taking an entirely obvious decision on the evacuation of Prypiat.

The evacuation of Chernobyl and the local villages was spread out over an even longer period: eight days. Until 2 May not one of the most senior leaders of the republic had been to the site of the accident.

Why did people, people entrusted with great power, with great privileges, but with a still greater moral responsibility, people who on solemn days of festivals and jubilees have become accustomed to being in view of everyone, why did they not share with their people its misfortune, why did those few kilometres that separate Kiev from Chernobyl prove so insuperable? Where does this moral harshness toward one's own countrymen come from?

A long time after Chernobyl there was a catastrophe in one of the Donbass mines. And the republican leader who went there, appearing on television, could not bring himself to utter any simple, human, sympathetic words about the great grief; he just announced that the mine was working at its 'normal work rhythm'... What has happened to us? When will we become people again?

Whatever the situation may have been, they learned faster in Moscow than in the capital of Ukraine that something very alarming and extraordinary was happening in Chernobyl, and they took the decisive and so necessary measures.

Only the visit to the Chernobyl area on 2 June by E. K.

Ligachev and N.I. Ryzhkov played a decisive role in the deployment of additional measures to contain and counter the scale of the accident.

Today we speak a great deal about the new thinking. It is not only the elected people, those who create international policy and so on, who must be convinced of it, but also those who are in the thick of everyday national life: the powers that be at all levels, and the ordinary citizens. This new thinking has to be educated from the school bench. And in its foundations we must place both profound specialist knowledge, the ability quickly to evaluate a situation and efficiently react to the changes it brings, and firm moral principles, the ability to defend our views without fear of those 'higher placed.'

10

One From the 'Little Football Team'

'In the photograph our team "All Stars" looks like an illustration of the growth of humanity from the times of Malthus up to our days. We are standing on a wooden platform over the sea, holding each other by the hand—the whole team, from the smallest to the biggest. The first is Maksym, the second Bondi, the third Iurek, the fourth Slavko, the fifth Ilko, the sixth Lionia, the seventh Lionia's father, who somehow wormed his way into our company, and the eighth is me (180 centimetres, 95 kilos). We are holding tight onto each other's hands, as if we are a living chain of generations and, it seems, no power will be able to tear us apart and separate us. The first one standing there is Maksym, and I am amazed why in the photograph he has no angel's wings. With my own eyes I could see those white wings edged with gold, hanging on the wall in his home; perhaps his father, afraid he might get entangled in wires—there were so many in the town—prohibited Maksym from using them. And perhaps there was some reason unknown to me. In any case, concerning Maksym's angelic origin I have no doubts at all. His thin body and fine neck are crowned by a large, high-browed head, with his hair trimmed under a flowerpot. In his face two great grey eyes shine, always shining, with benevolence for everything which surrounds him.'

I beg the reader's forgiveness for the quotation from my own work, but it was simply indispensable; it is an extract from my story *The Little Football Team*, written in 1970. The story was dedicated to the memory of the young Kiev poet Leonid Kyseliov, who died from acute leukemia, Almost all the heroes of this story are real people, although it was written in a grotesque-fantastic manner. And the little football team existed, Iurek, Bondi, Slavko, Ilko and, of course, Maksym.

Maksym Drach. The son of the distinguished Ukrainian poet Ivan Drach.

I have long known and loved Maksym and dare to assure the reader that I was in no way exaggerating when I described his angelic appearance and the features of his character. I must say that Maksym remained the same, in spite of all the changes in his voice and the fact that he grew up into a lean and lanky fellow, like a pole; he remained a very kind and very radiant fellow, although what sort of fellow is he now?

In 1974 Ivan Drach published a book, *The Root and the Crown*, containing a cycle of poems dedicated to the builders of the Chernobyl nuclear power station and the town of Prypiat. The main tone of these poems was optimistic, and quite naturally so: after all, Ivan Drach himself entered, no, not entered, but tore like a rocket into Ukrainian poetry as the prophet of new times, the powerful new rhythms of the epoch of the scientific and technical revolution. He has poems on genetics, cybernetics, and on nuclear physicists: the folkloric, profound songs which are the origins of Ukrainian poetry in some wonderful way join in him with a sharpened conception of that 'strange world' to which twentieth-century civilization rushed headlong. In the poem 'The Polissia Legend' the river Prypiat conducted a dialogue with the birds and the fish, fighting in alarm at their atomic neighbour. The river explained that for the Atom 'people build a castle of steel, and in some ten years throughout the world people will construct for it unshakeable atomic thrones.' Already in this, rather romantic conception of the construction of the Chernobyl nuclear power station there arose, through the lively tonality of the poem, a scarcely concealed alarm at the fate of Polissia nature, which had struck the poet with its primeval purity. Even more alarming and intuitively prophetic was another poem by Drach from this selfsame cycle: 'Mariia from Ukraine—No. 62276: From Auschwitz to the Chernobyl Power Station,' in which the poet told of Mariia Iaremivna Serdiuk, a builder of Prypiat, a person with a strange fate, a simple Ukrainian woman who went through the hell of Auschwitz and remained unshakeable in her kindness and love for people. 'Little woman's fate, you flew up like a phoenix over Auschwitz and flared up to illuminate the Atom-city over the Prypiat'—that is how the poet finished this poem. What vague, anxious sounds were born in his soul in those days, when in the land of rivers, sands and pine-trees the first contours of the nuclear station, hanging over the Kiev Sea and Kiev, were only hinted at? Could Ivan Drach have thought that his son, Maksym, would be forced to enter a struggle with the nuclear calamity of Chernobyl?

Maksym Ivanovych Drach, twenty-two years old, sixth-year student in the medical faculty of the Kiev Medical Institute:

'I first heard about the accident on Sunday morning, 27 April. I work in the resuscitation block of the cardiological centre in the October Revolution Hospital. I work as a medical assistant. A senior one, wherever they send you. I came on duty at nine in the morning. At half past nine one woman (her husband is an internal service major) said: "They've taken my husband off somewhere, like some nuclear station has blown up, but I think it's a joke." But at midday we were telephoned and told, in connection with the accident at the Chernobyl nuclear power station, to prepare, together with the general resuscitation department, forty beds on the fourth floor. Our block is on the second floor. I went up to the fourth to prepare the department. The patients were transferred to other departments, the beds were changed, all the medicines were prepared, blood substitutes and so on.

No one knew what we were going to have to deal with.

'At 6pm we were told that the first patients from Prypiat were already at the disinfection centre. We went to receive them: our duty doctors, from the department of radioactive isotopes diagnostics and general resuscitation. We looked at them. They were primarily young fellows: firemen and workers at the nuclear power plant. At first they went upstairs with their things, but then a frightened dosimeter operator-doctor rushed in and shouted: "What are you doing? You're 'glowing'!"

'They were all taken down, measured and led off to wash not in the disinfection centre but in the radioactive isotopes diagnostics department, where all the water was collected in containers and taken out. That was sensible, because that way the disinfection centre wasn't polluted. They were given our work pyjamas and, dressed like that, they were led upstairs.

'There were twenty-six in the first group. I went upstairs to them.

'We didn't ask them many questions; it wasn't the right moment. They were all complaining of headaches and feebleness. The headaches were such that one fine two-metre fellow was standing there banging his head against the wall and saying: "That's better, that way my head hurts less."

'Well, we at once began blood disinfection procedures on them, giving glucose transfusions. We put them all immediately on a drip, organizing it all well. I ran about between the block and the department. Since I've been working in the hospital three years I know everyone and where to get things, I've worked on the different systems, the different technical equipment, this and that. A student in the same year as me, Andrii Savryn, was there. He was also working in resuscitation, but general resuscitation. Normally he would not have been working that day, but he had just called in to pick up his photographic equipment. And he remained at work. If you have to, you have to. There were many doctors and staff there.

'The patients told us that the reactor was burning and had exploded, that they were loading sand, but what exactly was happening, there was no time to say, and the situation was such that conversation was not permitted.

'I was on duty till morning.

'The next day I went off, as usual, to lectures in the medical institute. On 1 May I was on duty again, but now in the cardioblock. I knew there were already more patients and that they were getting ready to free another floor for them. The eighth.

'On 2 May we were informed on television that E.K. Ligachev and N.I. Ryzhkov had come to the Chernobyl area, and I thought that since things were at such a level they just wouldn't be able to do without us, the medical students. I calculated purely analytically that it was far easier to get organized groups of students together in the hospitals than doctors. But on 4 May, in the morning, during the first lecture, our vice-dean came and told the lads to get ready—we were leaving at eleven o'clock. I went home, took a jacket, sweater, trousers, sneakers, cap and something to eat... They put us in a fancy bus, the Intourist type. The journey there was fine. But forty came back in a bus with eighteen places. Ok. We'll survive.

'We assembled in the medical institute in front of the exit by the radiology department. There they measured us all. At first we did not know what our work was to involve. They talked about work in permanent and field hospitals—to the extent of loading earth and digging trenches. I took two operating suits and masks just in case.

'We got into the bus, the mood was happy, we joked. Before leaving we were given potassium iodide. There one friend bawled at us: "Just when will these idlers be leaving?" The fact is, someone had poured that potassium iodide from a measuring glass out of the window onto his head, when the bus moved off. We really laughed.

'We came to Borodianka, the area hospital. We were distributed among villages and hospitals. One very distinguished medical chief from Moscow, a little tipsy, came up to us. He said what we would be doing, that today the evacuation of a thirty-kilometre zone was beginning. One of ours asked: "And what about the dry law?" He said: "Lads! There's no dry law in the adjoining areas. Drink as much as you can. Just so long as you can work. But remember that you are medical students and don't fall on your face in the mud. It's radioactive."

'They took us to the villages. From village to village, leaving us to reinforce the medical personnel. I ended up in Klavdiievo. I settled in in the hospital, in a ward. There were two of us, me and my friend Mykola Mykhalevych from Drohobych. We put our things down, it was already night, and set off. We stopped to check cars leaving the Chernobyl area. We had one fixed dosimeter, with a cable attached to the car, and two DP-5s working on batteries. We stayed there till about 2am, then the head doctor collected us and I slept till 6am. But at six he said: "Lads, one of you come with me." At work I've got used to getting up suddenly, so I said: "I'll come." We went off somewhere far away, on the road. I can remember a field, and in the field there were disinfection chambers, a fire engine, a table, and glasses and bread on the table. And ambulances, from Poltava and Zhytomyr.

'There we conducted a dosimeter check; we checked the background radiation in buses and on people's clothes.

'I worked there from 7am on 5 May till 10am on 6 May. A little over twenty-four hours.

'At first there wasn't much movement. Big military helicopters, in camouflage, flew over us; they flew very quickly. They flew low overhead; the noise filled my ears. The traffic on the road somehow throbbed. It was a long while from 10am to 1pm. Kiev buses were on the road, particularly "Icaruses," seventeen to twenty per convoy, and there were buses from Obukhiv and Novoukraiinka; all the places were familiar, that's why I remembered them.

'There were people in the buses. Basically from the village of Zalissia. It was twenty kilometres from Chernobyl. At that time not everyone left, because some of the people remained in the village, to load the cattle and domestic animals.'

I can remember how in those days, in a solid stream toward those who were going into the accident area, there came trucks loaded with cows. The animals stood there indifferently in the backs, sadly looking at the trees in blossom, the houses and plank fences whitewashed for the festival, the dazzling green grass and the springtime flood of the rivers. There were very complex problems connected with the decontamination of the cattle, insofar as the coat "took up" quite a lot of radioactive dust. And those cows that managed to get into meadow and feast on fresh grass also took in radioactive iodine and cesium. Animals like that were slaughtered at the meat-processing plants and their meat was put away in specially prepared refrigeration depositories, where it would gradually lose its radioactivity, designated iodine-131, an isotope with a short half-life.

M. Drach: 'In our brigades there were also female laboratory workers, they at once took blood from people for leucocytes. There were a lot of things in the buses, and we measured these things with the dosimeters.

'At first there were jams, then we adapted it so that the buses were let through in three rows, so there wasn't any disorder. One of us measured the bus itself, and two the people. The people got out of the bus, stood in a line and they came to me one at a time. Up to a certain level we were still letting people through. Where the level was higher, we sent them to get washed, to shake the dust off their things. There was one case where one grandad's boots were "radiating" a great deal. "But I washed my boots, lads," he said. "Off you go, grandad, you've got to shake some more off." He went, washed his boots, and his level was much lower. We sent him to wash three or four times.

'There was almost no one aged twenty to fifty. Why? It was said they had either run off (there were those who abandoned either their children or their parents), or they had stayed behind to work. So they were primarily old people, hunched old men and women, and little children. We measured the children's thyroid glands too. We had an order that when the thyroid gland radiated twice as much as the background, then the child had to be hospitalized. I didn't see any.

'Once they were through the check, the people got back in the bus. It was considered that they had been washed. They really did wash them. It's true, I did come across buses with a high level. Our lads caught a KamAZ bus—it was terrible the radiation it had. It was from Prypiat. That KamAZ was rushed straight into the field, six hundred metres away, and abandoned.

'So the whole day the convoys passed by. Toward evening they began to bring people's property. They brought the big bulky things separately. On Kovrovets tractors with trailers. We caught a dozen really dirty trailers, with dust-covered things. They were sent off to be washed.

'During the night we put a lamp on the table and sat there in our smocks. Isolated buses came by with people, catching up with their convoys. I remembered a Belarus tractor. In the cabin next to the driver was an old man, his father perhaps. The old man was carrying a hen and a dog. And he said: "Measure my dog." I said: "Grandad, shake your dog's hair well when you get to your destination."

'There was also one policeman, a young fellow in a little truck. He said: "Friend, measure me for radiation." I said: "Get out then, friend." And he said: "I can't get out, friend. I've done so much driving, picked up so much of that radiation that I can't get out of the cabin. I'll just get my feet out for you... " He hung his feet out, I measured—there was a lot! I said: "Friend, you must shake your boots."

'Later, when the evacuation was over, we did medical examinations and compared the data of blood analyses with other data. We took those who felt ill to the hospital for observation. I transferred these people.

'On 6 May we were brought protective clothing: black suits, caps, boots, gas masks. We were told that correspondents were on their way.

'But on 8 May we were sent to Kiev. A replacement came for us, men from the stomatological faculty.

'Well, on 10 May I went to lectures, as usual, and returned to work in the October Hospital. In May there were a lot of my type of patients, heart cases: obviously, the stress was making things difficult, we had a lot of work in the block.

'Around 11–12 May I noticed that I was sleeping a great deal but not feeling refreshed. I usually sleep five to six hours and feel fully refreshed. Now I was sleeping eight to twelve, even fourteen hours and not feeling rested. And I had become sort of "soft," lazy. A blood analysis was done and I was put on the eighth floor in our department.'

I can remember that ward and the eighth floor of the cardiological block, where the students of the Kiev Medical Institute who had been working on the effects of the accident were: Maksym Drach, Dima Piatak, Kost Lisovy, Kost Dakhno and Volodia Bulda. Professor Leonid Petrovych Kindzelsky and I came to the department. The professor consulted the students, examined their medical histories and studied the results of their blood analyses. Later Maksym Drach was to meet Doctor Gale, who visited Kiev at the beginning of June.

Now Maksym Drach and his friends are healthy, nothing threatens them. Their final examinations are ahead.

In my story *The Little Football Team* I foresaw for Maksym Drach the following future: 'I think that he will become a wandering philosopher, the Skovoroda of the twentieth century.'

I made a mistake.

Maksym, without doubt, will become a marvellous heart surgeon, considerate and sensitive, who, just like Skovoroda, will bring good to people, but a good strengthened by the new technical achievements of twentieth-century medicine. And, at the very beginning of his medical life, Maksym will have been trained by a unique experience, acquired in the days of the great national calamity, when he saw how complicated and conflictual everything is, how the high and the low are neighbours in the current of anxious events.

After the consultation we went out with Maksym to the spacious terrace-balcony of the cardiological block, which stands on a hill. From here there opened out an epic view over Kiev—the eternal city—spread out on the Dnieper-side springtime hills.

We stood, we looked, we thought.

What was happening in Kiev during those days?

11

The View Over Kiev

The scorching May of 1986 laid its new marks on Kiev: in those days the city, already a clean place, was washed and 'licked' to an incredible cleanliness. Incessantly, every day street-cleaning trucks prowled around the city, directing their watery whiskers and washing, from the hot asphalt, the dust in which radioactive nuclides lurked. At every entrance to a building, office, shop or even church there were wet cloths, and the endless cleaning of footwear became an unfailing sign of good form. The city streets were just as busy, but if you looked really closely you could notice that there were far fewer children: during the first days of May the city had rushed to get its children out, by whatever means—organized, disorganized, by train, by plane, by bus and by Zhiguli. To the west, south and east stretched out long convoys of cars with belongings on the roofs. Parents drove, taking the children, grandfathers and grandmothers out, taking them to relatives and acquaintances, and many to just anywhere, so long as it was as far as possible from the radiation.

In the press anouncements those days it was emphasized that Kiev and the Kiev area were living their normal life. Yes, people did not cower before the calamity, people struggled with the accident and its effects, and the outward appearance of the city hardly changed, and its inner, really tough essence was preserved, because there the businesses, transport, shops, institutes and offices all went on working, communication (true, with some interruptions) functioned, and the newspapers came out.

... It seemed to me in those days that I had never met so many pretty girls in the city, that until now there had never been such an enchanting spring there. I shall never forget how, arriving back from Chernobyl, I found myself in the evening twilight which had come down over Kiev. Everything was so normal: above the underground station 'Livoberezhna' the silhouette of an unfinished hotel skyscraper showed out dark. Opposite, at the taxi rank, cars' hoods sparkled, like a school of many-coloured fish which had come for the night to these sandy lands. An underground train hurtled toward the bridge and plunged into the thick of the Kiev hills and to rumble to the Khreshchatyk. Below the underground bridge the Dnieper overflowed and its expanses, hidden in the haze, had a Gogolian vastness and pathos about them. On the embankment courting couples kissed, and exhausted people headed for home—and all these simple, ordinary pictures of life suddenly got through to the depths of my soul, as if some illumination had reached me, the comprehension of some very important shift which had taken place in my consciousness over the last days. This peaceful evening seemed penetratingly beautiful to me, as if I had for ever bidden farewell to the spring, the city and life itself, and unfamiliar people had become close to me, and the everyday life of Kiev stood before me in a new light.

In the anxious light of the accident which had occurred so nearby—only two hours away by car—these were the days when a sense of danger had been sharpened to its limit. Later this vanished.

The Dnieper, the hills, the buildings and the people: all that was so everyday then seemed to me so unusual, as if it had come from the screen of some science-fiction film. In particular, Stanley Kramer's film 'On the Beach' frequently came to my mind in those days. It tells how, after the third and last atomic war in the history of humanity, Australia fatalistically awaits the appearance of the radioactive cloud. What seemed most strange and improbable in the film was that, in this critical situation, people just carried on living as they had always lived, without changing their habits and pleasures, maintaining their outer calm, existing as it were by inertia. It turned out that this was true. The habits of the people of Kiev remained those of earlier days.

However, the patriarchal, ancient city with its gold-topped cathedrals, preserving the memory of the ages, in about two weeks had changed unrecognizably, becoming closely united with the image of a new, atomic age. From being a resonant metaphor, in vain repeated by us before the accident, this word combination ('atomic age') was transformed into a severe reality. The words 'dosimeter check,' 'radiation,' 'decontamination,' all those 'millirems,' 'bers,' 'rads' and so on firmly entered the vocabulary of the people of Kiev, and the appearance of a man in a special suit, with a gas mask on his face and a Geiger counter in his hands flashed everywhere, became usual, just like the jams of cars at the exits from Kiev: at all the control points there were dosimeter checks for cars.

Milk and milk products disappeared from the stalls at the Kiev markets; the sale of salad, sorrel and spinach was prohibited. Other gifts of the Ukrainian earth: radishes and strawberries, new potatoes and onions, they were subject to dosimeter check. 'Honest, there's none of that radiation,' the peasant women vowed at the Bessarabka, as they sold strawberries at a knock-down price. But hardly anyone bought any.

And, as always happens, the children began to copy the incomprehensible life of the adults. On Rusanivka I had already seen children running through the bushes with a stick, as if they were measuring the background radiation with a dosimeter. They were playing 'radiation.' And one girl, wrapped up in a bedsheet, walked around the entrance to her building and, making 'terrifying' eyes, prophesies in a voice from beyond the grave: "Oo-oo, I am radiation, hide from me. I am evil and terrifying... ”

‘In Kiev there is a business-like, working atmosphere,’ the newspapers, radio and television assured us, and it was true. Ancient Kiev had preserved its appearance, its distinction both for its own sake, and for the country's sake, for the whole world's sake; visitors to the capital of Ukraine, surprised and respectful, would emphasize this.

It was all like that.

But in those days there existed another Kiev, hidden from the gaze of outsiders, a Kiev which did not attract the attention of the newspapers and television, and not to mention it would be to conceal a part of the truth, to distort the complex picture of events. It was a city of excited crowds around railway and airline ticket windows. There were days when even people who had tickets could not get through into the railway station. You had to get the police's assistance. The trains left with eight to ten people in four-seat compartments; speculators were charging up to a hundred roubles for a fifteen-rouble ticket to Moscow. I was almost moved to tears, although I'm not a very sentimental person, by Ievhen Lvovych Ierusalymsky, a candidate of medical sciences and senior scientific collaborator in the Kiev Oncological Institute, whom I had made the acquaintance of only three days before all this story. He came to me and offered me a ticket to Moscow for my daughter. And although the ticket was not needed, such a thought was in those days a sign of the most sincere friendship... Then, as during the war, a whole series of our usual conceptions changed in a trice. Such eternal notions as loyalty, decency and duty again acquired particular significance and value. In many Kiev apartments the telephone rang out in May from different parts of the Soviet Union. Friends, relatives, acquaintances phoned and invited people to come and stay. But there were people who did not phone, although it would have seemed, according to all the pre-Chernobyl laws of friendship, that they had a duty to do this.

For a long time, a whole month, I waited for a phone call from Moscow from one person whom I considered a real friend and who, over rather a long time, would come and stay with me. It didn't come. Then, quite unexpectedly, the Armenian writer Gevorg Mikhailovich Agadzhanian, whom I had only met once in my life in Kiev, telephoned from Baku. He was inviting my daughter for the summer...

We came to learn many strange and unexpected things in those days. What do you think, what were those long line-ups for in the department store at the beginning of May? For Finnish suits, for West German 'Salamander' shoes, or for Yugoslav leather jackets? No way. For suitcases and bags.

In those days Kiev apartments were literally bursting with conversations and rumours, arguments and prejudices, fictions and real facts. Decisions were taken and immediately withdrawn, fantastic projects were put forward, anecdotes and true stories were told over and over again. Stories spread persistently around the city about black Volgas driving up to the station square, about long lines for plane tickets in the ticket kiosks set up in some of the most notable buildings of the capital...

Yes, there was no panic in Kiev. But there was real concern for the health both of children and of adults, and it was worthwhile paying attention to this unease too.

Everyone remembers the photographs of the ruined reactor which spread around our newspapers. Even people who had no skill in atomic power engineering were stunned by the unnatural appearance of the reactor. It was clear to the specialists that something unprecedented in its sheer scale had happened. The first fallout went to the north-west and west. On 30 April the wind changed direction and blew toward Kiev. Radioactive aerosols were drawn toward the city with its many millions of inhabitants. I remember that day well; I was in the Ukrainian Health Ministry. I remember how anxiety and tension grew among the doctors, how in the ministry's offices and corridors there was talk of taking urgent preventive measures. Proposals rang out to turn to the population with a special appeal regarding security measures. Until 6 May no one paid any attention to these proposals.

Many people now blame the medical personnel: why didn't they warn us, why didn't they act earlier? I am not going to protect my colleagues, they have many sins on their conscience. But for the sake of fairness I want to stress that it is not the medical personnel who control the media channels. And the most important decisions are not taken by my medical personnel either. But decisions were very necessary. Already at the end of April it was really necessary to think seriously about the expediency of holding the May Day demonstrations in Kiev and the areas adjoining the Zone, particularly as regards the participation of children. I am certain that the love of Soviet people for May Day and their patriotic feelings would in no way have diminished because the demonstrations had been cancelled. I was told how in Belorussia they cancelled one of the first post-war May Day demonstration because of... rain. And what happened? And in 1986 too the people would have correctly understood the necessity of containing the accident and keeping children temporarily off the streets. And the people would have been grateful for this. Because photographs of the stricken reactor and of smiling children with flowers standing in holiday columns just do not bear any comparison. Was it really not possible to ask the people who filled the parks, beaches, suburban woods and went out to dachas on holidays temporarily to resist the charms of spring? People would have understood.

You can shake your head: the radiation level in Kiev didn't exceed the limits of acceptability. What, you say, is the purpose of cordoning off the city? But there are also limits of acceptability of alarm and worry, and those days they had exceeded all credible levels.

It was impossible and incorrect to ignore in those days the dread engendered by the radiation, and to struggle with it either with the help of silence or by means of lively and optimistic declarations. After all, for dozens of years the newspapers, radio, television and popular science magazines had themselves implanted, educated this dread, describing the horror of nuclear war, all its effects on the body and our genes. And although the scale of the Chernobyl accident and a nuclear explosion are simply incomparable, the dread of radiation turned out to be very great. And it would have been possible to play it down and to soften the psychological effects of the accident only in a situation where there had been a timely announcement of normal prophylactic measures—not on 6 May, but earlier. As the popular adage goes, 'God helps those who help themselves.'

As I wrote then and can now repeat even more sharply and persuasively: one of the most strict lessons of the first month (and of the following ones as well) of the 'Chernobyl era' was given to our mass media, which did not manage to restructure themselves in the spirit of the XVII Party Congress. The stormy development of events acutely shortened the time necessary for a shake-up, for all manner of agreements and liaising. Several of the most difficult days in our life come to mind: from 26 April to 6 May, when it was plain to see that there was a shortage of information in the country, and foreign radio stations were on the air, tormenting the souls of those who could receive them. Don't let us mitigate our shortcomings with lies. There were many people like that, because nature does not suffer emptiness, including an emptiness of information. And this inflicted not only an ideological but also an purely medical loss. Today it is already difficult to calculate how many people were, in those days, in a state of acute stress, in ignorance and in fear for the lives of their children and loved ones, for their health.

In Kiev 'catastrophists' made their appearance too, broadcasting all sorts of panicky false interpretations, and lively 'optimists' who could only repeat one message: all is fine and couldn't be better. In the city, in the heat of April, you could come across strange people wrapped up from tip to toe in old clothes, in caps, hats or head scarves, concealing almost half their faces, wearing gloves and stockings... These were the 'catastrophists,' who were mobilizing all means of individual defence. I don't condemn them, but after the Zone with its problems all the Kievan fears seemed simply ludicrous.

After the first days of silence, when information was very thin, there finally appeared numerous articles in the newspapers, and specialists began to appear on television... But...

In some publications and television programmes one perceived sort of falsely lively and frivolous intonations, as if this wasn't a great universal human tragedy, not one of the most awful events of the twentieth century, but a practice alert or firemen's competitions on models...

The habit of working according to the old schemes had had its effect, old schemes inherited from the times of universal equability, a desire to publish only lulling, joyous information; you felt there was a fear of extending glasnost to some very delicate and unacceptable questions, one of which was Chernobyl. Of course, it would be unfair not to notice what was new in those days in the work of the mass media. Let us take at least the interesting experiment undertaken by Ukrainian television: every day, beginning in May, the editors and operators of the popular information programme *Camera Today*, people who were not only talented but also daring (you have to agree, it was not simple to take pictures of things in the Zone, at risk from the radiation), gave the Ukrainian viewers details of what was happening at the nuclear power station and around it.

But this was already later.

And between 3 and 6 May dark rumours spread through Kiev: people said that an explosion at the station was imminent, because the temperature in the reactor had risen to the highest limits and the burning core of the reactor, if it went through the concrete casing, could come into contact with the water which had accumulated under the fourth block, and then... Some assured us (these were the 'catastrophists') that there would be a hydrogen explosion, others (the 'optimists') only an explosion of steam. There was nothing to be happy about in either case. People said that the evacuation of Kiev was being prepared, and they said a lot more too...

12

"The Danger of an Explosion Has Been Eliminated"

Strangest of all is that on this occasion the rumours were not without foundation.

From the press:

'Academician E. Velikhov said:

"The reactor has been damaged. Its heart is a red-hot active zone, it is, as it were, "on the brink." The reactor has been covered over with a layer of sand, lead, boron, clay, and this increases the weight pressing down on the structure. Below, in a special reservoir, there may be water... How will the red-hot core of the reactor behave? Will we manage to hold it or will it penetrate the earth? Never was anyone in the world in such a complex situation; it must be evaluated very precisely and no mistakes can be permitted... "

'The subsequent development of events showed that the direction chosen to confront the ruined reactor was correct' (*Pravda*, 13 May 1986).

From the article by V.F. Arapov, lieutenant-general, member of the Military Council, chief of the Political Administration of the Red Banner Kiev Military District:

'... The chairman of the Government Commission set a task for the commander of the special militarized company, communist and captain Petro Pavlovych Zborovsky:

'A critical situation has developed at the damaged reactor.

There is possibly water in a special reservoir underneath it. If the concrete base does not hold, irreparable damage may result. All must as quickly as possible find a correct solution and organize the pumping out of the water.

'. . . An armoured troop carrier took Captain Zborovsky and two volunteers, Junior Sergeant P. Avdeev and Corporal Iu. Korshunov, to the place from where it was necessary to penetrate the rooms leading to the reservoir. Radiation survey equipment showed the one could safely remain near the concrete wall only for twenty minutes. These daring fellows set to work in rotation. Then an opening was made, and Captain Zborovsky stepped into the unknown. Shortly after he proposed to the Government Commission a reliable means of pumping out the water, which was approved.' (*Raduga*, No. 10, 1986).

Mykola Mykhailovych Akimov, 30 years old, captain:

'It turned out that we had to work in a zone where there was very high radiation. So together with Captain Zborovsky (and with him there was Lieutenant Zlobin too) we decided first of all to take volunteers. When it was announced that eight volunteers were needed, all the personnel standing in the ranks took a step forward. Among them were Senior Sergeants Nanava and Oliinyk.

'We worked at night, by the light of lamps. We worked in protective suits. You have seen those suits—such a green colour, they're called 'ZZKs' (*Zahalnoviiskovyi zakhysnyi kostium*, protective suit common to all ranks—JIP). The situation which had developed at the station made it absolutely clear that we had to act quickly and decisively. The personnel understood the task we had been set, as is only proper, and at the station there were no superfluous instructions or clarifications. Just work.

'We worked in the Zone for just twenty-four minutes. In that time we had laid nearly 1.5 kilometres of pipe, set up a pumping station and begun to pump the water out. Everything seemed fine, we were pumping out the water. But, as people say, "misfortune doesn't walk alone."

'Hardly had we laid the pipe and begun the pumping than in the darkness of night someone's vehicle with caterpillar tracks crushed our hoses. They had been measuring something and in the darkness had not noticed the hoses. It was a lack of co-ordination. All this happened in a zone of high radiation levels. There was nothing for it. We got dressed and went back there. We went with another set of volunteers from our company. Water was streaming out under pressure, and the hoses couldn't take high pressure on them, so they leaked. And the water was radioactive. This flooding created an additional danger. We had to stop the flow immediately, and tighten the hoses where the water was gushing out. Altogether we plugged a lot of leaks in the hoses.

'What do I want to say about the lads? There are all sorts of things in our lives. As they say, there's no work without emotions. And when we got there and looked... No, at first we didn't feel any fear; well, we went in, and still ok. There were even birds flying around. But later, when the radiation readings began (each of us had his own individual dosimeter), when we understood that our organism had begun to take in roentgens, then the soldiers had a quite different attitude. I won't conceal anything: when the dosimeters started giving readings, then fear appeared. But none of the soldiers at the station showed weakness, they all carried out their tasks courageously, with high professional competence. There were no cowards among us.

'The task was set up outside the Zone. When we went into the Zone, there was no time to give orders: first, it was inconvenient, we were wearing gas masks, and secondly, giving commands just won't work, you have to do the job quickly. The lads didn't hesitate, I didn't notice them. They all knew they'd already taken up roentgens, but each one carried out his task.

'Apart from that, technical equipment is technical equipment. The pumping station was in a high radiation zone, working in a closed room, and it was practically impossible to stay there. But because of the shortage of air and the amount of gas the machine would stall. So from time to time—and this time amounted to approximately 25–30 minutes—we entered the Zone, ventilated the room, got the machine going again, and everything started off again.

'And this continued for twenty-four hours. We carried out this work during the night of 6–7 May. After that the pumping station was replaced.'

'Did you understand that this was one of the most important operations in the entire Chernobyl epic?'

'Yes, we understood that. Particularly the officers. We understood that if the water came into contact with that boiling mass, then there would either be an explosion or, in the extreme case, a vaporization... We understood everything. Everything we did we were fully aware of, we knew what we were in for.'

'Don't you regret choosing the profession of fireman?'

'No. I was born in Rostov area, in the village of Orlovske. It's the native land of Budenny. The steppes of the Salsk region. I graduated from the Kharkiv Firefighting Technical Academy of the Ministry of Internal Affairs, I was a "star pupil." I joined the army and I've been serving in Kiev for six years already. So you can consider me a Kievan. I don't regret my choice of profession, I made it consciously.'

'In those days all Kiev lived on dreadful rumours. Did you have the feeling that you had achieved something extraordinary?'

'You know, there was a feeling of relief that we had done our job. When we could report: "The danger of an explosion has been eliminated." We just didn't think that later we'd be interviewed. What we thought of was: "This soldier has taken up so many roentgens. He's got to rest. These will go first. They've taken up less."

'We looked after each other.

'And then it suddenly turned that we were sort of heroes. What I think is that everyone who was in Chernobyl did a job that had to be done. Everyone without exception. If it hadn't been us, then some other person would have been there in our place. We just went there as specialists.'

Besik Davidovich Nanava, 19 years old, senior sergeant:

'I was born in Georgia, in the town of Tskhakaia, and I grew up there. My father was an engineer, my mother an accountant. I've been serving for one-and-a-half years.

'What was it like? We were sitting in the club, watching a movie. The command suddenly came: "Firefighting company on alert!" We all immediately assembled and the company commander, Captain Akimov, said: "Lads, get ready, prepare yourselves for work." He gave us a briefing on security measures. When I heard all this, I thought of my home, of everything. But, you know, I felt it was necessary, that it was indispensable to do this. They've called us out, that means they need us.

'On the morning of 5 May we arrived in Chernobyl. We stayed there a whole day. On 6 May Major-General O.F. Suiatinov arrived, and the command came: our special operations group had to be at the power station. The company was fully prepared, and Captain Akimov said: "Volunteers, one step forward." Well, we don't need that... Every man took one clear step forward. Well, the healthiest, those best physically prepared, were chosen. I went in for sport, judo. We got the vehicles ready, checked the hoses, and on 6 May at 9pm we arrived at the power station. We arrived in armoured troop carriers. There were five officers there: Captain Zborovsky, Lieutenant Zlobin, Captain Akimov, Major Kotin and Major-General Suiatinov. And the eight of us—sergeants and soldiers.

'When we arrived, the major-general asked: "Shall we start at once or shall we have a smoke break?" Well, we talked this over: "We'll start right away." We didn't get out of the vehicles, but made our way to the work area. We arrived. We set up the pump and started to lay the hoses. At 2.30am we finished the work, came back, went through decontamination, had a wash and lay down to rest in the bunker. And at 5am the order came to go back. It looked like some reconnaissance vehicle with caterpillar tracks had cut through the hoses. There was contaminated water everywhere... All that work... We got up, got changed, and went to the accident area, changed the hoses, and back we came. All this took about twenty-five minutes. Three hours passed and all the time there was a helicopter on duty overhead, and from the helicopter they notified us that water was gushing out, there was a hole in the hose. It had to be dealt with immediately. So they got us up again. We went off there right away. Tightened it, fine. We were replaced at once and sent to the hospital for observation.

'Now I feel fine. I didn't write to my parents about this. But do you know what happened? They gave me leave, I arrived there, and my father looked at my military ticket, there where the radiation dose had been noted down. He asked me: "Son, where've you been, what's this?" Well, I didn't give any concrete explanation, but he knows about these things and guessed right away. He said: "Tell me what it was like." Well, I tried to soften it up a bit. I didn't want to give any natural picture, what it was really like. But they found it all out for themselves.'

The night of 6–7 May 1986 will enter history for ever as one of the most important victories over the stricken reactor. I don't want to seek out any agreeable symbols and get carried away by solemn comparisons. I've already got sufficiently carried away, and that'll do. But the symbols impose themselves: it happened on the eve of Victory Day. And now for me those two dates have become tightly bound in one knot. As long as I live, I always, during 'short May nights' will recall May 1945, ruined, burned, but triumphant Kiev: the Studebakers in the streets, the anti-aircraft guns in Shevchenko Park, being prepared for the holiday salute, the tears in the eyes of the adults. And alongside—May 1986: the armoured troop carriers rushing to the Zone, and the words of one officer who came to see us in the hospital: 'Lads, congratulations on your victory! There'll be no explosion!'

In the collective which carried out this responsible task for the Government Commission, I found myself together with veterans of the Great Fatherland War. The meeting had been organized by Stanislav Antonovych Shalatsky, an extremely interesting person, an experienced journalist, a colonel in the Soviet army and at the same time in the Polish army. At the end of 1944 he was editor of the newspaper of the First tank division of the Polish forces 'The Panzers,' precisely the one in which those such popular heroes of the TV film *Four Members of a Tank Crew and a Dog* had served.

Two Heroes of the Soviet Union came to this meeting. They were the ace pilot, Colonel Georgii Gordeevich Golubev, who during the war had served with the legendary Pokryshkin, and the illustrious intelligence operator, who saved ancient Cracow from destruction by the Hitlerites, Ievhen Stepanovych Berezniak, known to the whole country as 'Major Whirlwind.'

Colonel Golubev gave us a vivid and honest account of the difficult work of a fighter pilot, notably about the real work, not about the 'heroic exploits': the physical strain which weighed on the aces' shoulders, the various technical 'shortcuts' which pilots resorted to in war. If you don't shoot the enemy down, he'll shoot you down. And Berezniak told us of the intelligence operator's work behind enemy lines, when a man is in a constant state of tension, all the time carrying within himself an oppressive feeling of danger. In such circumstances the most daring, the most calm and ingenious are the ones who survive.

I looked at the young, 18 to 19-year-old shaven lads with red epaulettes on their shoulders and I saw how attentively they listened to the veterans' stories. I thought: in forty years these lads, with their grey hair, will recount in just the same way their burning days in Chernobyl and, in precisely the same way, with bated breath, the children of the twenty-first century will listen to them.

But if I had said this to the soldiers, they wouldn't have believed me, they would have laughed. For today they cannot imagine themselves as old men.

13

The Flight Over the Reactor

From the first days of the accident the situation around the stricken reactor was monitored. To achieve this all possible means were exploited, on the ground and in the air.

Nikolai Andreevich Volkozub, fifty-four years old, senior inspector, Air Force pilot of the Kiev Military District, pilot-gunner, colonel, Master of Helicopter Sport of the USSR:

'On the morning of 27 April I was told by telephone to go to headquarters with all my personal defence gear. It was a Sunday. A car came, I quickly got ready, arrived at headquarters and there found out what had happened.

'I received the order to fly to the town of Prypiat. When I flew past the power station, I involuntarily went to one side and saw the whole picture. Those places were familiar to me, I often flew there. We switched the helicopter's on-board dosimeter and straight away as we approached the nuclear power station we noticed that the radiation levels were rising. I saw the ventilation chimney, the ruined fourth power block. There was smoke, and in the middle you could see flames, in the ruins of the reactor. The smoke was grey.

'I arrived over Prypiat and heard the voice of our leader. So our leader was already there, Major-General Mykola Trokhymovych Antoshkin. I landed at the stadium. A car came up to me. I asked: “And where’s the pad?” They replied: “Near the flower-bed, by the Town Executive Committee.” I took off and landed near the flower-bed. I arrived in Prypiat around 4pm. The town had already been evacuated. Except for the cars parked by the Town Executive Committee, the town was empty. This was very unusual.

‘They were loading sacks near the river station and taking them straight to the central square. From there helicopters went to the reactor. At first they didn’t hang the sacks outside, but put them inside the helicopter. On the approach to the reactor they opened the doors and simply shoved the sacks out.

‘On 27 April our helicopters threw out sacks until darkness fell. At the Government Commission they reported that they threw out—I can’t remember precisely just now—I think it was a little over eighty sacks. The chairman of the Commission Borys Ievdokymovych Shcherbyna said that this was nothing, a drop in the ocean. It was very little, they needed tonnes.

‘We flew to the base and thought: what could we do? The question was put to everyone for discussion: flight officers and technicians. Throwing sacks out by hand was unproductive and dangerous. One flight technician: well, how much can he throw out? And during the night of 27–28 April everyone thought: how can we do this better? As you know, in theory the external attachments of an MI-8 can take two-and-a-half tonnes. And that night an idea took shape: the load should be hung on the external attachments. The sacks would be put in the brake parachutes of fighter planes—they’re very strong—and then hung underneath. On helicopters there are special attachments to suspend loads. You press a button, and it releases the load. And that’s that. At first we worked on MI-8s, then we got more powerful machines.

‘Our command point had been set up on the roof of the Hotel ‘Polissia’ in the centre of Prypiat. From there the power station was clearly visible. You could see how the helicopters, having left the pad, set a combat course to release the load, and you could direct them. A complication was that we did not have a special aiming device for releasing something hanging outside, that mass of sacks dangling under the machine. In working out a methodology for the flights, we established that the crew had to maintain a flight altitude of 200 metres. It couldn't be lower because of the radiation, and besides, the ventilation chimney was 140-150 metres high. This was very close. We had to aim for the chimney. That was our main reference point. I can see it all the time... It'll remain in my memory all my life, perhaps. I even know where which fragments lay on it, no one else saw them, but I had made them out. On the chimney there were platforms.

'We maintained a speed of eighty kilometres an hour. And our leader followed the flights with a theodolite. They designated a point, and when the helicopter came to a halt at that point, the command came: "Drop!" We worked it out that everything would fall directly into the ruins of the reactor. Later they placed another helicopter a little higher, and it checked how good the aim was. They took a photograph and at the end of the day we could see how precise we had been.

'Later we thought up another improvement: we fixed it so that the parachute stayed behind and only the sacks went down. We unhooked two ends of the parachute. And after that, when we worked in more powerful helicopters and dropped lead pigs, we dropped them using freight parachutes, designed to take down military equipment.

'A few days later we set up a pad in the village of Kopachi. It was also near the nuclear power station, but the radiation levels were lower.

'The fact that radiation has neither taste nor colour nor smell at first dulled our sense of danger. No one paid any attention, either to the dust, or to anything. We worked as much as we could. There were gas masks, but, you see, the soldiers, when they were loading the sacks, they pushed the gas masks up onto their foreheads, so they were like eyeglasses, and they worked...

'Later, when this came out, there were briefings and

medicine went to war, they began to punish people.

Afterward, when the wind turned toward Kopachi and the radiation levels rose steeply, we changed pad and went to Chernobyl.

'On these flights I trained crews, explained to them how to drop the load. To help us crews began to come from other units. We already had some experience, and every fresh crew received very detailed instructions from us. We elaborated diagrams of how to suspend the load, how to make the flight, and how to do the drop. Everything, absolutely everything. You do a briefing, a readiness check, you do a flight as co-pilot, do one more mission, and then you're ready to fly.

'After the flights there was a health check and a decontamination of the helicopters.

'On 7 May we stopped filling in the reactor. As soon as we had stopped, at one of the meetings of the Government Commission the scientists and specialists decided that in order to work out the next steps in eliminating the accident, they had to know the temperature in the reactor and the composition of the gases being emitted. At that moment it was impossible to get near the reactor on foot or in a vehicle because of the high radiation levels. One of the scientists, academician Legasov, proposed doing it from a helicopter.

'No one had ever before carried out such a task. What made it complicated? A helicopter, through its aerodynamic properties, can hover over the land either at a height of around ten metres (this is called hovering in the danger zone) or higher than 500 metres. From ten to 200 metres is a prohibited zone. Why? The helicopter is in general a reliable machine. I've been flying them since 1960. For me it's like a bicycle. Whatever the situation, even if an engine refuses, I will always land. But if the helicopter is hovering at a height of up to 200 metres and an engine refuses, the pilot, however high his qualifications, will not land the helicopter, because he won't be able to get the rotor into the autorotation, that is, gliding, mode. But this is only when he's hovering. If, however, he's flying horizontally, then everything's fine. The helicopter can go into the autorotation mode only above a height of 500 metres.

'So one of the dangers is hovering higher than ten metres. This is prohibited. Only in extreme situations can it be permitted. Secondly, the emission of heat from the reactor. No one knew its thermal characteristics. And in a zone of increased heat the power of engines decreases. Well, there were increased radiation levels too. And one more thing: the crew cannot see what is happening underneath.

'Everyone understood these complications. But there was no other solution. Everything was measured by wartime standards. But the measurement had to be made. The task consisted in lowering the active part of the equipment, that which indicated the temperature, what is known as a thermocouple, into the reactor.

'An Air Force commander flew in and set us the task. He said: "It's a very complex task. But it has to be done. How can we do it?"

'He asked me. I said: "Yes, it's complex, but we have to try. We'll do some training." I have a good deal of experience, I've flown all types of helicopter, so, clearly, they had the idea of charging me with this.

'The training started. I immediately outlined a plan of how to do it. At that time I had switched myself off completely from everything else, concentrating my attention only on this flight. Apart from the co-pilot and the flight technician, the doctor of technical sciences Evgenii Petrovich Riazantsev had to fly with me. He was deputy director of the I.V. Kurchatov Atomic Energy Institute. Evgenii Petrovich explained to me that a thermocouple was a metallic tube on a cable. Oleksandr Stepanovych Tsykalo, relief chief of the dosimeter operators, was to fly with us too. These people, together with whom I had to work in complex circumstances, will be remembered.

'We had to think of how to lower it, this thermocouple, into the reactor. I went off to our engineers and said: "Put your engineers' thinking caps on and let's think." Although I had some ideas too.

'We took a 300-metre cable. You know, it's a bad stimulus, an accident, but if we always lived and worked like then, with such efficiency, without fuss, when everyone gave all his strength to the job, we would have a completely different life... In literally half an hour the cable was ready.

'The wire from the thermocouple was twisted around the cable. A weight was hung onto the end of it. The cable was laid out at the airfield. I chose the helicopter myself, so that it was a powerful one, and I tried the engines. I set the task out to the crew. I didn't notice any excitement, but it just had to be done. I calculated how much fuel to take. I didn't need any extra, so they just gave me a little bit over what I needed. I took off and flew up to the cable. I hooked it on and flew directly off. I raised it. On the ground they had made a circle of about the radius of the reactor, twelve to fourteen metres, using parachutes. I started to imitate the flight. A load of around 200 kilos hung below me. I came in smoothly, hovered, reduced speed, and slowly approached the circle. Our leader corrected my bearings. I hovered. And he gave the command: "Stay right here." I picked out a reference point for myself so that I hovered just right, and locked myself on, but I knew intuitively that I was hovering in the exact spot. I held it there. But he said: "You're hovering just right, but the load is swinging like a pendulum." I was hovering at a height of 350 metres, but the load was swaying.

'I hovered there for five minutes, it swayed; I hovered for ten minutes, still swaying. It wouldn't settle down. I hovered there and the thought came: "What can I do?" This training too is quite dangerous, even if morally easier: there's no radiation and raised temperature. But from the point of view of aerodynamics, it's dangerous. But you don't think of this when you're flying.

'I could see there was nothing doing. I came to the landing place and lowered the cable onto the pad. I unhooked it. Then I landed.

'And an idea came to me: what if, along the whole length of the cable weights are hung at certain intervals? It should stabilize. We strung lead bars onto the cable. Our engineers did it all efficiently.

'This was all done on 7 May, at night.

'On the next day I flew out to train with this cable. The

cable took the strain well. I began to descend. I just barely touched the ground with the cable (I heard the command: "Touching!"). I moved off and the cable stood there like a pillar... In these situations piloting needs the technique of a jeweller... I did it once more to convince myself that it was possible. If only you could have seen our expressions after that flight. In general a flight with something suspended outside is considered one of the most complex... After that I did a few more practices.

'On 8 May we were brought the thermocouple. It was like a wire. The end is a sensor. We connected everything up and laid the cable out on the pad in Chernobyl.

'On 9 May Evgenii Petrovich Riazantsev and Oleksandr Stepanovych Tsykalo came. They set up the apparatus in the helicopter. Just before the flight we ourselves, the crew, using lead sheets, made a protective shield: we laid it on the seats and the floor. But not there where the pedals and feet were, that was impossible. We covered everything up well. We were given lead waistcoats. We explained to our passengers how the flight would go, and we covered them too with sheets and agreed on how everything would be coordinated. My colleague Colonel Liubomyr Volodymyrovych Mymka, positioned on the Hotel "Polissia," was going to watch over the flight.

'Well, we got into the helicopter and took off from Chernobyl without any problems. The end of the cable had been marked with an orange ring to make it more visible.

'I approached at a height of 350 metres. I had to find out what the situation was there with the temperature and the power of the engines. The helicopter hovered solidly.

'The flight leader said to me: "Fifty metres to the building... Forty... Twenty... " He prompted me with the height and the distance. But when I was right over the reactor, neither I nor the leader could see any more whether I had hit the mark or not. So they sent me another MI-26. It was piloted by Colonel Chichkov. He hovered at a distance of two to three kilometres behind me and could see everything. I had to hover to the side of the chimney...

'And Evgenii Petrovich Riazantsev was himself looking through the hatch and telling me where I was with gestures: "Over the reactor." We made temperature measurements at a height of fifty metres, forty, twenty and in the reactor itself. Evgenii Petrovich saw everything. And the equipment was noting it down. When it had all been done, I came away.

'Behind Prypiat a special place had been designated, and I dropped the cable in some sand. The cable was radioactive.

'Six minutes and twenty seconds had passed from the moment we had been hovering. It had seemed an eternity.

'It was a victory.

'On the following day, 10 May, we were given another task: to determine the composition of the gases being emitted. Again the same thing, the same cable, but not a thermocouple, this time a container on the end. This task was easier: we didn't have to hover, but just pass smoothly over the reactor. On 12 May it all had to be repeated with the thermocouple. By then we had experience, and some calm. But in spite of the fact that we had, it would have seemed, some experience, there was no way we could hover for less than six minutes.

'You have to approach, stabilize the cable, then you begin to come down and take the measurements. How did I feel? From 27 April we hadn't had a single peaceful night, we slept for two or three hours. And we flew from dawn until dusk. People often ask me: "What effect does radiation have?" I don't know what it does, but the exhaustion was dreadful, and where did it come from? Was it from the radiation or from the lack of sleep, from the physical exhaustion or from the moral and psychological tension? There really was tension, it was a great responsibility.

'After these three flights I continued flying, carrying out radiation reconnaissance.

'Altogether I hovered for nineteen minutes and forty seconds over the reactor.'

From the press:

'With the aim of reducing the radioactive fallout, a protective shield of sand, clay, boron, dolomite, limestone and lead is being created over the active zone. The upper part of the reactor has been covered by a layer which is composed of more than four thousand tonnes of these protective materials.' (From a statement by the chairman of the Government Commission, deputy president of the Council of Ministers of the USSR B.Ie. Shcherbyna at a press conference for Soviet and foreign journalists, 6 May 1986, *Pravda*, 7 May 1986.)

'Professor M. Rosen (director of the nuclear safety department of the International Atomic Energy Agency) responded positively to the methods applied by Soviet specialists to absorb the radiation, using a shield made up of sand, boron, clay, dolomite and lead... Work continues under the damaged block, in order completely to neutralize the radiation fire and, as the physicists say, "bury" it in a thick concrete case.' (*Pravda*, 10 May 1986.)

'From the Council of Ministers of the USSR. On 10 May at the Chernobyl nuclear power station work continued to eliminate the effects of the accident. As a result of the measures taken, the temperature within the reactor has fallen significantly. In the opinion of scientists and specialists, this testifies to the virtual halt in the combustion process in the reactor's graphite.'

14

Dr. Hammer and Dr. Gale

From the press:

'On 15 May M.S. Gorbachev received in the Kremlin the eminent American entrepreneur and public figure A. Hammer and Dr. Gale. He expressed deep gratitude for the sympathy offered by them, for their understanding and rapid concrete assistance in connection with the calamity which had afflicted the Soviet people, the accident at the Chernobyl nuclear power station... In A. Hammer's and R. Gale's act, emphasized M.S. Gorbachev, Soviet people note an example of how relations between the two great peoples should be built at a time of the existence of political wisdom and goodwill in the leadership of both countries.' (*Pravda*, 16 May 1986.)

On the morning of 23 July there landed, at Kiev's Boryspil airport, a white Boeing-727 with the flag of the United States on the fuselage and a blue and red inscription on the fin: 'N1OXY,' which denotes number one in the company 'Occidental Petroleum Corporation,' whose president is Armand Hammer. On this aircraft, which is provided with every indispensable item, from a study to a bathroom, the indefatigable 88-year-old businessman 'clocks up' hundreds of thousands of kilometres, directing the complex and many-sided operations of the Occidental company.

Armand Hammer, with his wife and also Dr. Robert Gale with his wife and three children, had arrived in Kiev.

And immediately after their arrival Hammer and his retinue set off for the cardiological block of the Kiev October Revolution Clinical Hospital No. 14. The very hospital where Maksym Drach works in the resuscitation block. Putting on a white smock and recalling his medical youth (after all, by education he is a doctor), Dr. Hammer went around the sections in which, after the accident at the Chernobyl nuclear power station, over 200 people, who had been in the danger zone, were kept for observation. That day in the department there were only five people, called in for planned repeat observation.

Dr. Hammer took a sympathetic interest in the general state of each of them. He was assisted by Dr. Gale, who, having been in Kiev before, had already examined these patients.

On that very same day Hammer and Gale flew over the fourth reactor in a helicopter. I flew with them, and now, whether asleep or awake, what I saw gives me no peace: the flight over the fourth reactor, hovering over the massive, white and dead structure of the nuclear power station, swallowed up by the half-shade, over the white and red striped chimney, over the mirror-like surface of the dead cooling pond, over the winding riverbed of the Prypiat, over the eccentric interlacing of wires, supports, over the heap of auxiliary structures, and over the abandoned technical equipment. As in every memory, hidden in the depths of time, the real forms progressively wear away, many things lose their clear outlines, but the feeling of anxiety and pain remains immutable, just as it was during that summer as evening approached. Pressing against the windows, we, the passengers in an MI-8, studiously peered into a magical picture, which attracted our eyes: the black nozzle of the fourth reactor, the ruined structures, the debris around the base.

After the flight around the fourth reactor, standing before movie and television cameras, Armand Hammer said:

'I have just returned from Chernobyl. What I saw made such an impression on me that it is difficult for me to speak. I saw a whole town—fifty thousand inhabitants—and there was not a single person around. Empty. Buildings, big buildings, all empty. Here and there even washing hangs out, the people didn't have time to take it in. I observed the work being done to save the reactor, so that there would be no more problems with it. I would like every human being to visit here, to see what I have seen. Then no one would talk about nuclear weapons. Then everyone would know that that is suicide for the whole world, and everyone would understand that we have a duty to destroy nuclear weapons. I hope that when Mr. Gorbachev meets Mr. Reagan, he will tell him everything and show him a film about Chernobyl. And later, in the future, when Mr. Reagan comes to Russia, I would like him to visit Kiev and Chernobyl. Let him see what I have seen. Then, I think, he will never speak of nuclear weapons.'

He is a surprising man, Armand Hammer. Perhaps the secret of his imperishable alertness is that he knows how to relax. After our helicopter rose above Kiev, he immediately fell into a doze. Dr. Gale considerately covered him with a white cape. But no sooner did the word Chernobyl resound than this wise old man was transformed, as it were, sharp-sightedly looking into the green landscape beneath us, over which the shadow of our helicopter peacefully crawled, like a ghostly mowing machine. He noticed everything, even the sixteen-storeyed buildings in Prypiat, even the washing on the balconies; all of it stiffened and unnatural. And on the way back he fell asleep again.

In the evening of that same day Armand Hammer flew out from Kiev for Los Angeles.

But Dr. Gale stayed behind with his family for several days, to meet Kiev colleagues, to rest in our city, and to familiarize himself with its monuments and museums. After all, during his first visit to Kiev on 3 June Dr. Gale had no time for that: he was to consult with a group of patients who had been under treatment in the Kiev X-ray-radiological and Oncological Institute, in Professor L.P. Kindzelsky's department.

I accompanied Dr. Gale during his first visit to this institute. Dr. Robert Peter Gale looks younger than his forty years (every morning an hour of obligatory jogging); he is tanned, concentrated and taciturn, and the look of his grey eyes carefully and searchingly fixes on the person to whom he is talking. In spite of his outward coldness and his typically American businesslike manner, he is a very likeable person and it is pleasant to have dealings with him; he replies so considerately and patiently, and in such an informed manner, to the numerous questions put to him by correspondents. And he is elegant too. He invariably wears a dark-blue blazer with gold-like buttons, a dark-red tie, and grey trousers. And his bare heels at first look somehow very ridiculous and moving: he wears shoes without heel-pieces. It appears that this is a Los Angeles habit, walking 'barefoot': in Gale's native land it is always warm.

Just before we entered, we all—the guest and his companions—put on white smocks, caps and masks, and tied protective slippers on our feet. And suddenly we became surprisingly similar to one another; you couldn't make out who was the American, who the Muscovite and who the Kievan. A family of doctors, united by the common interest in saving humanity.

I saw how considerately Gale examined the patients, how he asked the casualties and the doctors questions, thoughtfully studied the charts with the data of analyses, interrogated on the finer points of the methods applied by the Kiev doctors. He was particularly interested in the cases of bone-marrow transplant.

And in this there is nothing surprising. After all, Gale is a well known specialist in the field of bone-marrow transplants, a professor of the University of California, the director of a clinic, the president of the International Bone Marrow Transplant Organization. Kiev professor Iu.O.

Hrynevych reminded Gale that he had visited his Californian clinic. Gale, on that occasion having heard out the assistants, who had just showed him a patient, after some thoughts clearly and confidently dictated a course of treatment and, raising his hands, said: 'God help us.' Gale smiled, remembering that meeting, and his severe face suddenly became childishly animated. Looking at the Kiev patients, saved from their serious situation, he superstitiously touched wood: even if it doesn't help, it won't do any harm.

Later, to my question: 'What do you believe in?, Dr. Gale gave a very serious answer:

'In God. And in science.'

Then, in those anxious June days, his visit to Kiev had been very short, and precious minutes ticked by in conversations with the press. In July Dr. Gale felt far freer: on the day after the departure of Hammer, the American doctor together with his wife Tamar, his three-year-old son Elan and his daughters—seven-year-old Shir and nine-year-old Tal—went to the Kiev Pediatrics, Midwifery and Gynecology Institute, where the director of the institute, Academician of the Academy of Medical Sciences of the USSR O.M. Lukianova, president of the Ukrainian section of the international organization 'World Physicians for the Prevention of Nuclear War,' met him.

Here, in this, perhaps, most sacred place in the world, the place where new life comes into being, where the struggle for the survival of humankind is waged, Dr. Gale's children very quickly made friends with the little patients, without feeling any linguistic or ideological obstacles. They exchanged presents and together sang the song 'Let there always be sunlight.' Then little Tal played the violin, and blue-eyed Shir regretted there was no piano; she too would have demonstrated her skill...

And Dr. Gale in the meantime held a professional dialogue with the pediatricians, midwives and heart surgeons. In the resuscitation department we stood for a long time near the plastic incubation units, plugged in to complex equipment: here tiny creatures lay, the future people of the twenty-first century, still unaware of the nuclear alerts worrying us today.

The Museum of the Great Patriotic War, the Museum of the National Architecture and Life of the UkrSSR in the village of Pyrohovo, the V.I. Lenin Museum—all were on Gale's schedule. The different stages in our history, the different sides of our life . . .

In the V.I. Lenin Museum a symbolic sculpture attracted Dr. Gale's attention: a little monkey, sitting on Darwin's *On the Origin of Species by Means of Natural Selection*, examines a human skull. The story of this sculpture is interesting. During his second visit to Moscow Armand Hammer gave Lenin this sculpture, which he had acquired in London. People relate that Vladimir Ilich, taking the gift, said: 'Here is what can happen to mankind if it continues to perfect and increase instruments of destruction. There will only be monkeys left on Earth.'

That was the prophetic warning of our Leader.

I retain many recordings of conversations with Dr. Gale, who, in fact, was very interested in literature and is himself the author of a publicistic book. I have tried to select the main ones from among these recordings:

'Dr. Gale, what led you to medicine? Was it chance or a conscious choice?'

'I made a conscious decision. In our society the profession of doctor is one of the most respected. I wanted to become a doctor.'

'How old were you when you made that decision?'

'I entered college when I was sixteen.'

'Was medicine a traditional profession for your family?'

'No, in my family there were no medical people. My father is a businessman.'

'Are you satisfied with your choice of profession?'

'A lot of people ask me: "Now that you have achieved international prominence, what do you intend to change in your life?" I always answer that I was entirely satisfied with my life even before I became well known and that I don't intend to change anything in my life.'

'Dr. Gale, I know many oncologists and many hematologists and I know that psychologically this is a very difficult profession. After all, a doctor constantly sees death and misfortune. What is your attitude to that?'

'You're partly right, Dr. Shcherbak. Psychologically this is a difficult profession. But, on the other hand, that increases its attraction. After all, it's a vocation. The oncologist and the hematologist must very often solve extremely complex problems and be in difficult situations, often because our knowledge in this field is limited. So it seems to me that oncology really gives a great deal of space for medical creativity. In college we often argued: which is better, writing music or playing music? If you are a cardiologist, you "play music." But in oncology "music is written." There everything is new and everything is unknown.

'Moreover, I have been trained both as a scientific worker and as a doctor. It is precisely in oncology and hematology that it is very easy to tie the results of laboratory research with work in hospitals, with the real treatment of the patient. After all, it is not a coincidence that the first illnesses for which their genetic nature was known were precisely illnesses of the blood: the destruction of the synthesis of hemoglobin, let's say. And, you know, the majority of Nobel Prizes in the field of medicine have recently been awarded precisely for the examination of these problems.'

'In connection with what we have just been saying: what do you feel more like, a doctor or a scientist? Are you in favour of a synthesis?'

'Being a good doctor, curing people, that is work which must take up all one's time. Even more. Being a real scientist, that too is for more than all your life. Sometimes it seems to me that no one can do the two in parallel. Particularly in our age, when both medicine and science have become so technological, let's say "technicapacious." And I am aware that there really just aren't enough people who could unite these two occupations. This is extraordinarily important. In my opinion, there has to be a synthesis. It's precisely in this that I see my duty: to unite in myself the doctor and the scientist.'

'How do you divide up your time during a normal work situation in your California clinic?'

'As director of a clinic I spend the greater part of my time on the rounds, examining patients, talking to them. My patients often have quite common forms of cancer, for example, lung cancer. And I am anxious for them, like any ordinary doctor. Some time goes on directing a small research institution which collects statistical data on the results of the application of new methods of the treatment of leukemias, the transplant of bone marrow and other information. And, finally, the most important thing I'm occupied with: my own laboratory, where fundamental research is done on the study of the molecular mechanisms involved in the development of leukemias.

'I realize that this sounds as if I'm dissipating my attention, but I don't think this is the case. I have concentrated on three directions, because before us there is a very important objective; we want an effective cure for leukemia. And we think that the first result will be obtained in a laboratory.

'Where are we going? What is the fundamental idea of our research? No child should die of leukemia. To achieve this we must do everything in our power.'

'Are there cases of cures in your clinic? Do you succeed in transforming acute leukemia into chronic leukemia?'

'In 1986 we succeeded in curing approximately seventy per cent of the children who had developed leukemia. And nearly thirty per cent of adults. If we do a general calculation, then we succeed in curing exactly half the patients.'

'That's a phenomenal result!'

'Unfortunately, the overwhelming majority of the population does not fully realize how far we have got in the treatment of leukemias. However, one half of our patients just isn't enough. After all, the other half die. For example, this year 200,000 Americans will die of cancer... '

'In the press it was stated that you have a Ph.D. What problem did you examine in your thesis?'

'My theme was life and death. The unity of life and death on the philosophical level. In my autobiography, which came out in the USA, I touch on this theme.'

'Dr. Gale, what do you say to your patients when you have made a diagnosis?'

'I always tell my patients the truth, I give them all the facts. I don't know if this is good or bad, but we subscribe to the philosophy according to which a person must have all the information. The fact is that the most important decisions regarding treatment are taken by the patient himself. And to do this he must have true information. This doesn't always work out in the best way, but we simply don't have a better way out.'

'Did you have anything to do with radiation sickness before you came to Moscow and began to treat the casualties of the Chernobyl nuclear power station accident?'

'Yes, we had a certain experience. Some cases of leukemia require bone-marrow transplants. And in such cases we give patients heavy doses of radiation, sometimes almost lethal doses. We have considerable experience in the treatment of patients radiated by massive doses of radiation, nearly several thousand *bers* (biological equivalent of roentgen).'

'Did your prognosis regarding the treatment of the patients in Moscow meet with the actual results?'

'On the whole, yes, if you mean the general pattern, the statistical prognoses. But it is very difficult to make a precise diagnosis in every individual case. In general, making a prognosis is a difficult ethical problem and a heavy burden. In the first place I have in mind the treatment of patients with leukemias in my clinic, not the treatment of the Chernobyl patients. We assume, I know, that out of a hundred patients who need a bone-marrow transplant, fifty per cent will survive and be cured. But this is no comfort for those fifty per cent who will die. Our treatment shortens their life. And so every time that a patient dies, a patient whose life was shortened by our treatment, I feel personally responsible. I have to be responsible for their death, but I have no other advice to give.

'On the whole it would be simplest not to do any transplants. But in that case we will deny the great majority of our patients the right to life.'

'Dr. Gale, whom do you remember best of your Moscow patients?'

'I must say directly that I remember every one of them. I remember each one as a person, as an individual. And some people left a very profound trace. I particularly remember three patients.

'The first is the doctor who worked near the reactor, giving assistance to the casualties. As a doctor, he was aware of the whole danger of the situation, he understood everything, but he conducted himself courageously. The second patient is a fireman. When I first left Moscow for Kiev—do you remember, at the beginning of June?—I was away from the clinic for three days. When I came back, he was very angry and asked me: "Where were you? Why did you go away?" And the third is also a fireman. It's possible he didn't realize what danger hung over him; it's possible he understood, but perhaps did everything particularly so as not to pay attention to the threat to his life. His behaviour was very moving; during the rounds he would always ask: "How are things, doctor, how do you feel?"

'Two of those patients died, the other survived... '

'What feelings guided you when you decided to come to the Soviet Union?'

'In the first place I am a doctor and I was aware of the possible effects of such an accident. So I considered it necessary to offer my assistance. As a representative of the medical profession, political rivalries do not concern me. Our first duty is to save people, to help them. Moreover, similar accidents can happen not only in the USSR, but in the USA and other countries. And it's natural that we shall expect the same sympathy and the same help on the part of Soviet people.'

'Do you think it is possible to draw an analogy between Dr. Hammer's visit to our country in 1921 and your present journey?'

'In a sense, yes. It's true that Hammer was then concerned with the problems of combating typhus, and we are combating the nuclear threat. The circumstances are quite different, but the essence is one and the same. In this sense nothing has changed. But the situations, categorically, are absolutely incomparable. Just think: just as the very idea of a nuclear reactor accident in 1921 was absolutely impossible, so today it is impossible to imagine a typhus epidemic on the same scale as in 1921. Humanity has learned to conquer all the difficulties on its path... '

'But this creates new problems.'

'It will always be like that,' laughs Dr. Gale. 'And today it is difficult for us to imagine what problems will worry humanity in sixty years.'

'This time you brought your children. Does this mean that nothing threatens their stay here?'

'Many people in the world think that Kiev has been totally abandoned by its inhabitants or that every single child has been evacuated. And one of the reasons which forced me to come here with my family was a wish to emphasize once more that the situation is perfectly under control and that the patients have been given the necessary assistance. I had no doubts about the safety of my stay in Kiev. And not at any price would I have brought my children if there had been even the slightest potential danger. I think that such an action is easier to understand than a whole series of medical statements and complex generalizations.'

'Do you consider that the situation in Kiev is getting better?'

'Of course. The radiation levels are going firmly down. But there are some things which require particular attention. For example, the problem of protecting the water. But all measures are being taken to protect the city of Kiev. For example, artesian wells have been bored, and alternative sources of water supply have been designated. I consider that the situation is under full control. In these questions I have full confidence in my Soviet colleagues. I am sure that they would not expose their children and themselves to levels of radiation which they considered dangerous.'

'Were you satisfied with the information you were given?'

'During my first visit to the Soviet Union, in particular to Kiev, I was always struck how sincerely and openly we worked together with our Soviet colleagues. I want in particular to emphasize that the communication of the Politburo of the USSR on the investigation of the causes of the accident at the Chernobyl nuclear power station made a great impression on many of us. I consider that the evaluation of the accident was surprisingly sincere. Probably even more sincere and open than we thought, and that pleases me greatly. I hope, indeed, I am convinced, that your analysis of the medical information will be as complete and open as the analysis of the physical causes of the accident.'

'Would you like to come to Kiev again?'

'Not only do I want to come, but I definitely will come. I shall return to your city in October, when the exhibition of works from Dr. Hammer's collection opens.'

Robert Gale kept his word. It was autumn, it was the same airport, it was an American airplane, only it was a little smaller than the 'Boeing,' and on its fin was the number '2OXY.' Together with Dr. Gale came the the popular American singer and composer John Denver, who performs his ballads in the 'country' style. On behalf of Dr. Hammer, Dr. Gale opened the exhibition 'Five Centuries of Masterpieces.' At the opening ceremony he said:

'Chernobyl has become for all of us a reminder of the fact that the world must for ever put an end to any possibility of the outbreak of a nuclear war.'

. . . That same evening there was a concert in the palace 'Ukraina,' all the money from which went into the fund to help Chernobyl. John Denver's words on his visit to the Piskarev Cemetery in Leningrad rang out sincerely and touchingly; after his visit he wrote a song about the power of spirit, the courage of the Soviet people, their love for their land. With great emotion the hall listened to the pure voice of this red-haired lad from Colorado. 'I want you all to know that I respect and love the Soviet people,' said John Denver.

'For me it's very important to be here, in the Soviet Union, and to sing for you. And not just to sing, but to share my music with you.. I want you all to know that I deeply respect the inhabitants of Kiev and the inhabitants of Chernobyl, I respect their bravery, their courage.' John Denver was applauded not only by thousands of Kievans, but also by Dr. Gale and his wife. And later there was a farewell party, a little sad, as always when you part from good friends. And when night fell over the city, we all went out together onto the bank of the Dnieper and sang to our American friends our national song, 'The wide Dnieper roars and moans.' Both Gale and Denver listened attentively, and then Denver thoughfully asked: 'And where is Chernobyl?'

We pointed to the north, into the darkness, there from where the Dnieper brought its autumn waters.

15

By What Are People Tested?

As I listened to the melodious and very compassionate songs of John Denver, I thought of Vladimir Vysotsky. An autumn day in Kiev in 1968 came to mind. The leaves were falling from the apple trees in the famous garden of Oleksandr Dovzhenko, in the film studio bearing his name. I was walking around near the 'Shchors' pavilion, waiting for Vysotsky. I had seen him in the film *Vertical* and it had seemed to me that I would certainly recognize him straight away. But when this small, skinny chestnut-haired fellow without a beard and wearing a leather jacket turned up, looking far younger than the hero of *Vertical*, I only realized at the last moment that it was him. And I realized this because a guitar was hanging over his shoulder. Those days the film *Quarantine* was being shot with my script; it told of how a group of doctors in a scientific research laboratory became infected with a dangerously contagious virus; in the film we strove to investigate the characteristics of the people, to model their behaviour in an extreme situation. To a significant extent the film's subject was contrived, even fantastic, but the doctors' characters were taken from nature. Vysotsky had agreed to write a song for our film, and the director, S. Tsybulnyk, who was not in Kiev at that time, had charged me with this song.

We exchanged a few words and went to the pavilion, where everything was already prepared for the recording. And when Vysotsky began to sing his song, I suddenly understood why, in ordinary life, it was difficult to recognize him: the feeling of watching someone 'monumental,' which you feel from his screen heroes, was created by his inimitable hoarsish voice and by his powerful temperament. A miracle of transformation took place simply before my eyes, as soon as the the first chords rang out on the guitar. I really liked the song, and we immediately took it into the film. It was performed by the fine actor and singer Iurii Kamorny, who later died tragically... But the recording made by Vysotsky remains in my cassette. Here is that song:

The volleys of arms have long since grown silent,
Above us, just the light of the sun.
By what are people tested,
If there is no longer any war?
And you often happen to hear
Now, as then:
'Would you go reconnoitre with him,
No or yes?'

The armour-piercing gun will rumble no more,
There'll be no funeral note pushed under the door.
And it seems—it's all so calm.
And there's nowhere to show yourself now.
And you often happen to hear
Now, as then:
'Would you go reconnoitre with him,
No or yes?'

The calm is only a dream, I know.
Get ready, steady, and fight.
There's a leader on peace,
Misfortune, danger, and risk.
And you often happen to hear
Now, as then:

'Would you go reconnoitre with him,
No or yes?'

In the fields the mines have been defused,
But we're not in a field full of flowers.
The search for the stars and the depths,
Watch you don't rule it out of the game.
And so we often hear,
If misfortune comes:
'Would you go reconnoitre with him,
No or yes?'

During the events in Chernobyl I often recalled this courageous song and the question asked in it: 'By what are people tested, if there is no longer any war?'

L. Kovalevska: 'On 8 May we left the village in Polissia for Kiev, to Boryspil airport. I had sent my mother and children off to Tiumen. I had little money left, and what I had I distributed to our people of Prypiat at the airport. Three roubles to some, two to others. The women with children were crying, I was sad for them. I kept a rouble for myself, to get to Kiev. The ticket from Boryspil to Kiev costs eighty kopecks, so twenty kopecks were left in my pocket. I was all 'dirty,' my pants were contaminated. I stood at the taxi rank and telephoned acquaintances: one was not at home, another had gone off somewhere. There was one address left. I thought: I'll take a taxi, get there, and I'll tell the taxi driver my friends will pay for me. And if they're not at home, I'll note down his details and settle up later. I stood there. Some man comes up, stands in line behind me and asks: 'What time is it?' You know the situation, men come up and ask, just to get acquainted. I stood there angry, unwashed, my hair a mess. I looked at his arm: was there a watch there or not? There wasn't. Then I told him what time it was. I don't know why, but everyone immediately guessed when you were from Chernobyl. Because people hardly know Prypiat, everybody would say: 'Chernobyl.' Whether it was the eyes or the clothes, I don't know. But they unfailingly guessed.

And so that fellow standing behind me in the line asked me: "You, are you from Chernobyl?" And I said angrily to him: "What, can you tell?" "Yes, I can tell. Where are you going?" I answered: "I don't know, I'm afraid it's a waste of time my going there." And he asked: "Well, have you got somewhere to spend the night or not?" "No." He took me by the arm and said: "Let's go." "I'm not going anywhere with you," I answered. You know, I thought: this man will take me to his place and all the rest. . . I know these things. No. He sat with me in the taxi and took me to the hotel 'Moscow.' He paid for the taxi, he paid for the hotel. Then he took me to his place of work, an old dear was on duty, he gave me a good feed and brought me back. I got washed, straightened myself up, and then I found out his name: Oleksandr Serhiiovych Slavuta. He worked in the republican society of bibliophiles.'

A. Perkovska: 'At the beginning of May we began to take children out to the pioneer camps. The things I saw! People knew there would be places in 'Artek' and 'Moloda gvardiia.' The parents began to come along. They put pressure on me so that their child went without fail to 'Artek.' Well, I gave these parents a good talking to, I won't hide the fact. But I often had to take a risk and its consequences. Our instructions were to send to the camps those children who had finished the second to the ninth grades inclusive. So people came up to me and said: 'What about those in the tenth grade, aren't they children too? And what are we to do with children in the first grade?' Just imagine: a mother comes up, she's alone, has no husband, has shift work and a six-year-old child. What's the child to do, wait till it's finished the second grade? What's the mother to do with it? Well, I just wrote in another date of birth without any twinge of conscience. Then, when I got to the pioneer camps, I was really rebuked. But look, what else could I do?

'So we drew up the lists and then it began. Kievans phoned in and asked me to take their children to the camp. And the like. When I began to look through the lists, I found all manner of forgery there. We had to have it announced on the radio that parents should come with their passports and show their Prypiat residence permit...

'In August I went out to 'Artek' and 'Moloda gvardiia'; I took some children. And can you imagine: I found an almost grown-up girl from another town. She hadn't the slightest connection with Prypiat. I even found a child from the Poltava area. How these children ended up in 'Artek' and 'Moloda gvardiia' I don't know. But they, like all the children, were there for two periods...

'When at the beginning of May I took some pregnant women to Bila Tserkva, a grandee—the third secretary of the Town Committee of the Party—came out and said: 'One must think like the state.' And they themselves met our women wearing anti-plague suits, gas masks and taking a dosimeter along the streets. And here in this very same Bila Tserkva they wouldn't receive the children until evening, because there was no dosimeter operator.

'And when I was resting in Alushta after the hospital, a friend of mine warned me: "Don't say where you're from. Say you're from Stavropol. It'll be better that way." I didn't believe her. Moreover, it's degrading to conceal who you are and where you're from. Two girls, from Tula and Kharkiv, sat down at my table. The asked me: "Where are you from?" "Prypiat." They immediately ran off. Then some "friends in misfortune" came and sat with me—women from Chernihiv.'

A. Esaulov: 'In our town, in the communications centre, on 29 April our telephonist Nadiia Myskevych fainted with exhaustion. She had been on the lines all the time. Our centre head Liudmyla Petrivna Sirenko was a fine person too, she was the first in the town to organize special work units. On one occasion, a lunatic cut off the power at a substation. He said: "I have symptoms of radiation sickness. Get me out, or else I'll switch off the power supply." He went and switched it off. So Liudmyla Petrivna immediately went over to the emergency power supply. That's a Woman with a capital "W."

'Here's another case. The deputy director of the nuclear

station dealing with social questions came to me and said: "Help me, Aleksandr Iurievich, I need to bury Shashenok, the operator who died in the fourth block. He needs to be put in a coffin and buried, but Varyvoda from construction administration won't supply a bus. It's the only one he's got." It's difficult to say in such a case who's right and who's wrong. He only has a single bus and he needs it for the living to solve some questions of life and death.

'We went to Varyvoda. I said: "Look, why are acting the fool? You've got to accord a man our last respects. Let's have the bus." And he said: "You can't have it." I said: "What are you, a parasite, you don't obey Soviet authority?" And he continued: 'All the same you can't have it. You can do what you like to me, but you can't have it."

'So I went out into the road, stopped the first bus I saw, handed it over to Varyvoda, and took his bus for the funeral... '

Iu. Dobrenko: 'After the evacuation nearly 5,000 inhabitants remained behind in Prypiat: people who remained on the instructions of various organizations to see some work through. But there were those who didn't accept the evacuation and stayed in the town, so to speak, illegally. They were primarily pensioners. It was difficult for them, for a long time we were still getting them out. I took one pensioner out on 20 May. It was an old grandad who had lots of awards, a participant in the Battle of Stalingrad. How had he lived? He went to the military, took a few gas masks off them, and even slept with them on. He didn't put the light on, so it wasn't noticed at night. He had biscuits and he saved water. When I took him out, the water in the town had already been cut off, it was necessary for the decontamination work. There was power, and he'd been watching the television.

'This is how we found him. His evacuated son came and said: "My father's stayed behind in the town. I said nothing for a long time, but I know that there's no longer any water in the town and that he's just sitting there. Let's go and get him." We got there and he said: "Ok, there's no water, I'll come." He put on a gas mask and took some buckwheat to make some soup or other. In the villages there were also a lot of old men and women like him, who for nothing in the world would leave their homes. We called them "partisans." It's true, they weren't all in the same situation. There were those whom their children had simply left behind. They just didn't take them. But they didn't make a song and dance about it, it was just "Stay on here and look after the house and property".'

Sofiia Fedorivna Horska, headteacher of school No. 5 in the town of Prypiat: 'Not all the teachers stood the test that fell to our lot. Not all of them. Because not every one turned out to be a pedagogue, a professional. When we had been evacuated, some left their classes and abandoned their children. The children reacted very badly to this. Especially the older children, those who were finishing school this year. The teachers who fled, abandoning their children, explain this by the fact that they had no experience, they didn't know how to act in such a situation, what to do. After they heard on the television that everything was back to normal, they reappeared. For us this was an instructive lesson for the training of future teachers. Those whom we select from among the pupils and for two years train to enter the pedagogical institute. Among the teachers there were "activists" who spoke louder than others but then fled. Yes, there were people like that.'

Valerii Vukolovych Holubenko, military director of secondary school No. 4 in the town of Prypiat: 'When the evacuation was carried out, we took neither the school journals nor anything. After all, we were to be gone for a short time, we hoped to return soon to the town. But the academic year was coming to an end and we had to write school-leaving certificates for the tenth-graders. There were no journals, so we proposed to them that they could put down their own marks. We said: you can remember your marks. When we were able to check, no one had raised their marks, and some had even lowered them.'

Mariia Kyrylivna Holubenko, headteacher of secondary school No. 4 in the town of Prypiat: 'When we had all been evacuated here, to Poliske, I was appointed a member of the parcels commission attached to our Prypiat Town Executive Committee. What really struck me was the kindness of our people, a kindness we felt just physically, when we opened parcels, sorted presents and read letters. Some of the things we hand over to old people's homes, there where the lonely old people of Prypiat now live, others go to mother and child buildings, and others to pioneer camps, particularly clothes for the smallest children. We get a lot of books—we hand them over to the libraries for the special construction and operational units of the nuclear power station. And here, in the next room, are nearly two hundred parcels and another three hundred are lying in the central Kiev post office. Many letters come from children. Leningrad children sent a lot of parcels with books, children's clothes, dolls and stationery, and in every parcel there is a letter, and in every letter there is anxiety and care. Although these children are third- or second-graders and live far from the place where the accident occurred, they have understood that this is a calamity. There are many parcels from Uzbekistan and Kazakhstan; they give dried figs, dried fruit, peanuts, sugar and tea; pensioners send soap, towels and bedclothes, and the children most often send books, dolls and games.'

But I ask the reader not to be too taken up by the gentle and moving emotions which welled up, perhaps, under the influence of the story about the parcels and letters sent by kind, decent and sincere people. We mustn't feel cosy. Because the events in Chernobyl engendered other things: the traditional masterpieces of our dimwittedness and red tape ridiculed by Saltykov-Shchedrin.

Here is one of them:

'The Yalta Town Council of National Delegates of the Crimean Area. 16.10.86. To the President of the Executive Committee of the Prypiat Town Council of National Delegates comrade Voloshko V.I.

'In accordance with the directive of the USSR Ministry of Health No. 110 of 6 September 1986 the Executive Committee of the Yalta Town Council of National Delegates decreed on 26.09.86 No. 362 (1) to grant an apartment in the Crimean Area to citizen Miroshnychenko M.M., a family of four people (self, wife, two children), evacuated from the zone of the Chernobyl Nuclear Power Station. We ask you to send to us certification of the surrender by citizen Miroshnychenko M.M. of a three-room apartment No. 68 with all conveniences and living space of 41.4 square metres in building No. 7 on Heroes of Stalingrad Street, the town of Prypiat, to the town authorities.

'Deputy President of the Town Executive Committee P.H. Roman.'

Witty, aren't they? All the country knows how and to whom the inhabitants of Prypiat (see 'The Evacuation') 'surrendered' their apartments with all conveniences. And only in sunny Yalta do they think that some ill-intentioned people or some relatives of the above-mentioned citizen immediately settled in the apartment abandoned on 27 April 1986 by comrade Miroshnychenko, with its 41.4 square metres of living space and, against established order, existing conditions and high levels of radiation.

Indeed, 'By what are people tested?'

The flash over the Chernobyl nuclear station with its blinding glow lit up good and evil, reason and folly, sincerity and pharisaism, sympathy and schadenfreude, truth and lies, disinterestedness and covetousness—all the human virtues and vices which lurked in the souls both of our countrymen and of those who lived far beyond the frontiers of our country.

I recall the May issues of the popular American magazines *US News and World Report* and *Newsweek*: ominous purple colours on the covers, the hammer and sickle, the symbol of the atom, and black smoke over the whole world. Bawling headlines: 'Nightmare in Russia,' 'Deadly Fallout from Chernobyl,' 'The Chernobyl Cloud. How the Kremlin Described it and the Actual Risk,' 'Chernobyl: New Health Worries. Perilous *guided tour* of Kiev.' And the first apocalyptically triumphant words of the articles: 'This was the unseen dread of the twentieth century... ' I admit that these sensational headlines and hysterical tone are in the tradition of the American press, which strives to get through to the reader and attract his attention at all costs. All this is normal. But, even bearing all this in mind, it was impossible to see in all this material the slightest human sympathy for those who had suffered in the accident, and behind the ominous medico-genetic prophesies one could sense not a shadow of anxiety for the lives and health of the children of Prypiat and Chernobyl. I was particularly struck by the cold politicking tone of Felicity Barringer's article in the *New York Times* (5 June 1986); this woman (woman!), with the sensitivity of a robot manipulating a pen, as if it were a scalpel cutting through living flesh, had a report from the pioneer camp 'Artek,' where the children from Prypiat were then holidaying. In her words there was no real female, maternal charity, just a hateful propagandistic incomprehension of what the eleven- and twelve-year-old children were saying to her, children stunned by what had happened, longing for the homes to which they would never return...

Later, in the foreign editing room of Radio Kiev I was shown dozens of letters which had come during those days from the USA and Great Britain. And I thought how much higher the ordinary people of these countries, and of our country, stood above these primitive propagandistic stereotypes.

In July 1986 an unusual present from the USA arrived at the Chernobyl fire station, where 'grandad' Khmel and his comrades had been working in April: a plaque with a message from the 28th Fire Unit of the town of Schenectady in the state of New York, on behalf of 170,000 members of the association of firemen of the USA and Canada. Here is the message:

'The fireman. He is often the first to arrive where there is danger. So it was also in Chernobyl on 26 April 1986. We, the firemen of the town of Schenectady, New York state, are moved by the courage of our brothers in Chernobyl and are deeply grieved by the losses which they have suffered. A particular comradeship exists betweenn the firemen of the whole world, people who respond to the call of duty with extraordinary courage and bravery.'

The vice-president of the International Association of Firemen James MacGovern from New York, and Captain Armand Capulo from the town of Schenectady handed this message on behalf of all decent Americans—and they stressed, that is the majority—to the Soviet representatives in New York. They spoke of our people with great respect. They recalled the principle professed by the decent people of the whole world: they sympathize with and help those who have fallen into misfortune, and they do everything they can to save them as quickly as possible from their misfortune.

... To Vladimir Vysotsky's question, 'By what are people tested if there is no longer any war?,' it was possible to give a unanimous answer in 1986: people are tested by their attitude to Chernobyl.

It is a pity that Vladimir Vysotsky was no longer among us, that his sorrowful and courageous songs about Chernobyl were not born. About those who went into the fire. Vysotsky was really needed by us, in the Zone.

16

The Last Warning

Exactly one hundred years ago, on 2 June 1887, staying in the Roslavl district of Smolensk province, approximately three hundred kilometres from Chernobyl, Vladimir Ivanovich Vernadsky, later an eminent Soviet scientist and academician, and the first president of the Ukrainian Academy of Sciences, wrote to his wife:

'The observations of Ørsted, Ampère and Lents gave impetus to the study of electromagnetism, which has incomparably increased the power of man and in the future promises radically to change the whole shape of his life. All this has its origin in observation of the particular properties of magnetite. And the question occurs to me: do other minerals not have similar properties... and if they do, then will they not reveal to us a whole series of new powers, will they not give us opportunities to apply them in new ways and to increase tenfold the power of mankind? Is it not possible to arouse unknown, terrible powers in various substances... ?'

This quotation was taken from I.I. Mochalov's very interesting article 'The first warnings of the threat of nuclear omnicide: Pierre Curie and V.I. Vernadsky,' published in the third issue of the journal *Questions on the History of the Natural Sciences and Engineering* for 1983. Omnicide is a comparatively new term, denoting the universal destruction of people.

In the letter written by a young, 24-year-old graduate of the physics and mathematics faculty of St. Petersburg University ten years before the discovery of radioactivity by H. Becquerel we have, perhaps, the first warning in the history of humanity of the approach of a new era, that era which has affected us so painfully in Chernobyl, proclaiming a complete destruction of humanity (omnicide) in the event of the wartime use of nuclear power.

All his life V.I. Vernadsky was concerned by an at first vague, and later more and more explicit vision of the use of this terrible power:

'We, the children of the twentieth century, have grown accustomed, with every step, to the power of steam and of electricity, we know how profoundly they have changed and continue to change the whole social structure of human society. And now before us are discovered in the phenomenon of radioactivity the sources of atomic energy, which exceed by a million times all those sources of powers which human imagination depicted to itself. Slowly, trembling and expectant, we turn our eyes to the new power being revealed to human consciousness. What does it announce to us in its future development?.. With hope and dread we peer at the new defender and ally.' (1910)

'Radium is the source of energy. It is powerful and in a way which we do not yet understand acts on the organism, creating around us and in us ourselves some changes which are incomprehensible, but wonderful in their effects... You feel somehow strange when you see these new forms of matter, obtained by the genius of man from the bowels of the Earth. These are the first small seeds of the future. What will happen when we are able to obtain them in any quantity?' (1911)

And so in those days, when dosimeter operators still went around Kiev and when the question of the appropriateness of total defoliation of the famous Kiev chestnut trees and poplars was being seriously discussed, I arrived at the building where Vladimir Ivanovich Vernadsky worked in 1919-1921. On the facade of the presidium of the UkrSSR Academy of Sciences there hangs a plaque commemorating this man of genius. It seemed that he came to the window of the president's study and carefully looked at us from the depths of the Kievan past, when drivers clattered past this building over the paved road and still few were those in the world who had heard the word 'radiation.' And no one took seriously the prophesies of the scientists.

I had come to see the vice-president of the UkrSSR Academy of Sciences, the eminent Soviet botanist and ecologist, academician **Kostiantyn Merkuriiovych Sytnyk**. Here is what he said:

'It is a tragedy, a great tragedy of nations, which has directly affected hundreds of thousands of people. A new ecological factor has appeared. I would not exaggerate it, but it would be far worse to underestimate it. Of course, it cannot be accepted that we, carried away by discussion of the Chernobyl problem, forget that the factories of Ukraine continue this very day to smoke, that the pollution of the Dnieper water basin by chemical and metallurgical works continues. However, this new factor, linked to the accident, exists, and it is a negative factor.

'People are very worried by its existence, and that is natural. The overwhelming majority of the population was never interested in what the maximum permitted norms of oxide of nitrogen and sulphuric anhydride were. But they are very interested today by the level of gamma-, beta-, and alpha-radiation. This may be explained by the fact that for years we talked of the tragedy of Hiroshima and Nagasaki and were told in detail of the massive danger for humanity connected with radiation. People gradually built all this up in their consciousness and now see radioactivity as a high-risk factor. We have here a certain psychological phenomenon, a certain split between emotions and knowledge. Everyone knows that as a result of industrial waste entering the environment there is an incidence of carcinogenic substances, but this does not create any particular excitement.

'Radioactivity is another thing. The people's mood is very anxious, after all, they fear for their children and their grandchildren, because we have said a great deal about the genetic, far-off effects. Both scientists and the mass media have a duty to pay attention to this.

'We must objectively and judiciously explain the existing situation, without evading people's anxious questions. We must not be afraid that this will create panic, because the reason for panic is precisely in a shortage of information... And we, like parrots, repeat one and the same thing, that products are unadulterated, that they have been checked, and so on. But I myself am not sure of this, and if I myself for several months do not drink milk, then how can I assure people of the opposite? Go to the station and look what the people are bringing from Moscow? Bags full of produce. Most of them do not trust what we write.

'Let us say: medical personnel in their over-optimistic communications in June and July repeated one and the same thing: that it was possible to bathe in the Dnieper around Kiev. At that time I considered that it was not at all possible to bathe. Because by the shore, in the silt, there had at that time accumulated a certain quantity of radionuclides, which have now settled on the bottom. Nothing would have happened to the people of Kiev if they had restrained themselves from bathing for one year or had not gone to the woods for mushrooms.

'At the same time, clearly, we must not exaggerate this problem. Why? Well, because in nature there is a powerful process of diluting and scattering radionuclides, and this saves us. Once again, for the nth time, Mother-Nature has become our saviour. I have in mind the trees, the earth and the waters of the Kiev Reservoir, which received and took up the basic radioactive fallout. We may have cursed the Kiev Reservoir many times, its threat hanging over our city, but in the present situation it has turned out to be a useful accumulator, taking into the silt a part of the radionuclides, which then settled on the bottom. The Reservoir turned out to be capable of taking a great deal of the radioactivity, it swallowed up a certain quantity of the radionuclides, and we hope that finally there will be a dilution of the radionuclides to tiny concentrations...

'The water problem is closer to me, because I head a working group on monitoring the state of the water in the Dnieper basin. The Dnieper is an important element in all our worries, perhaps the most important. After all, the water of the Dnieper basin is used by the thirty-five million people of Ukraine. Immediately after the accident a series of urgent measures were taken to protect the sources of the water supply, and I can say that the population of Ukraine receives drinking water of good quality. I can say that with a full sense of responsibility.

'At the same time we had to be ready for the unexpected. With this objective, we, together with the V.M. Hlushkov Institute of Cybernetics, made a mathematical model to study and make prognoses about the state of the water in the Dnieper basin. In this model we foresaw the various—even the most extreme—possible situations, and elaborated, in case of their arising, a whole complex of special measures. But so far no extreme situations have arisen.

'What are the lessons to be learned from Chernobyl? Not so very long ago we had a typical scientific conference on the problems of Chernobyl and its effects. No less than one hundred people assembled, with figures, charts and communications. Physicists, biologists, geneticists. There were interesting reports, some very interesting ones among them. This was not that optimism of which Chingiz Aitmatov writes in "The Block": "How long will we assure people that even our catastrophes are the best?" No, in our circle we are very open. Nonetheless, some objective data do give us grounds for optimism. But you need to be able to talk about this in such a way that people believe you. You need to find scientists who speak persuasively, with facts and figures, to arouse the faith of readers and viewers.

'And, of course, one of the main lessons is the moral lesson. As a result of the Chernobyl accident a feeling of bitterness and of disenchantment with science has grown much stronger. After all, you also spoke of this at the congress of Ukrainian writers, didn't you?'

'Yes, I did.'

'It isn't so much a question of science as of the moral qualities of individual scientists. You can very often observe the following situation: there are two or three scientists of approximately the same rank and title. One of them gives a categorical 'no,' and the two others 'yes.' What are those who decide to do? Naturally, they choose the answer which appeals to them more. Unfortunately, not even the scientist who says 'no' tries later to defend his view, to struggle for the truth, to talk in important forums, and so on. Even he avoids spiritual discomfort, does not want to enter into conflict with influential people and departments.'

Probably the main lesson of Chernobyl is that any, even the slightest blunder made by a scientist, any compromises with one's conscience must be severely punished. Because we have forgotten that once we did not offer our hand to a dishonest man. Once. And today the responsibility of scientists for their own discoveries and for the reliability of huge new constructions has increased a thousandfold. The scientist has a duty to fight tooth and nail for his ideas and convictions. But do you often see that?

Conversations like that were held in the rooms honoured by the name of V.I. Vernadsky, who said in 1922:

'The scientist is not a machine and not a soldier in the army, who carries out commands without thinking and without understanding to what they will lead and why they are given... To work on atomic energy it is indispensable to be aware of one's responsibility for what is discovered. I would like this moral element in scientific work, which, it seems, is so far from the spiritual elements of the human personality, like the question of atoms, to be a conscious element.'

The paths opened by Chernobyl brought me to Moscow too, where, forty years ago, on 25 December 1946, the first uranium-graphite atomic reactor F-1 'Physics 1' began its work. Then this was the outskirts of Moscow, Pokrovsko-

Streshnevo, and a dense pine-wood stood here. And the pines are still here. Now we have here the territory of the I.V. Kurchatov Institute of Atomic Energy.

I had come to see Valerii Alekseevich Legasov, academician, member of the presidium of the USSR Academy of Sciences, first deputy director and director of the department of the Institute, laureate of the Lenin and State prizes of the USSR. The principal scientific interests of Valerii Alekseevich are connected with nuclear technology and hydrogen power engineering, the chemistry of plasma and the synthesis of compounds of noble gases. But in 1986 the name of Academician Legasov resounded in the whole world in connection with the elimination of the accident at the Chernobyl nuclear power station. Valerii Alekseevich came to Prypiat on the first day after the accident and was appointed a member of the Government Commission.

I became acquainted with Academician Legasov shortly before I met him. While working on the popular science film *The Introduction* (from the 'Kievnaukfilm' studios), I, sitting in the clipping room, dozens of times ran through the film of the interview which Valerii Alekseevich had given to our filming group. I was particularly affected by the following words: 'I would like to draw attention to the fact that over many years this illness—insufficient attention to what is new—the inability to show what is new—was neglected, and it is not not all that simple to cure it. It has been neglected because, from our childhood, we are not seriously taught to evaluate what is new and to distinguish the old and the new. If you go into any old class and listen to the lesson, then, whether it is a humanities or a scientific class, as a rule you will find a situation where the pupils are being told what a fine book this is, what a precise equation this is, what a wonderful experiment this is. Never will you hear the question: How would you do it better, what is bad about this experiment, or how is this book unsuccessful?

'And by denying what seems good and ideal, creativity begins, the striving to do everything somehow better. Our school teaches us quickly to use what is available, rather than to throw out what has been achieved and create something new.'

This opinion seemed very interesting to me, but it revealed one of the reasons behind much of our confusion, particularly that at Chernobyl. Because our school directs all its efforts to educating obedient, well-behaved and industrious boys and girls, little 'yes-people,' and does not educate in them a spirit of criticism and an objective, with a 'for' and an 'against' approach to natural phenomena and social reality. It implants a normative way of thinking, and the child learns criticism (more often, despair and cynicism) from the street, sometimes relatives, acquaintances and books. Often the schoolchild has to break through to this on his own.

It was very interesting talking with **Valerii Alekseevich Legasov** about the lessons of the accident at the Chernobyl nuclear power station:

'It so happened that even before the Chernobyl accident I had to deal with questions of industrial safety, and particularly safety at nuclear power stations. In connection with Israel's bombing of an Iraqi nuclear research centre there was discussion of the effects of a possible attack on a nuclear power station not only in scientific, but also in the broadest circles. Our article published in the journal *Nature* (V.A. Legasov, L.P. Feoktistov, I.I. Kuzmin, 'Nuclear Power Engineering and International Safety,' (No. 6, 1985) was dedicated to this. Already then, examining this problem, we came to the conclusion that waging war when there was a high concentration of nuclear power stations was madness. Massive regions would have remained contaminated with radiation for a long time.

'But another question appeared to every sound-thinking person: what if we dispensed with nuclear power stations altogether? And in their place put their equivalents using gas, coal or oil? And so we started to think, and I repeat that this was before the Chernobyl events: suppose that a bomb hits a nuclear power station. That is bad. And what if it hits not a nuclear power station but a thermal one which has been built instead. And we saw that that too was bad. Explosions, fires, the formation of poisonous compounds will destroy a great number of people and render significant regions useless, even if for a shorter period.

'And after this you arrive at the opinion: it's not a question of the variety of power systems, but of their scale and concentration. The level of concentration of power of industrial sites is today such that the destruction of these sites, accidental or intentional, would lead to very serious consequences. In its development humanity has created such a compact network of different power sources, and different potentially dangerous component parts—biological, chemical or nuclear—that their conscious or accidental destruction would cause considerable trouble.

'The problem today is the proliferation of all sorts of sites and the concentration of vast power. At one time a restricted quantity of nuclear sites was brought into operation, and their safety was guaranteed by the high level of qualifications of the staff and the assiduous adherence to all the technological rules. Just outside this window works our first Soviet reactor, and it works well. But later, when reliable technical decisions yielded good results, they began to be applied on a large scale, and the productive capacity of the sites was increased as well. But the approach to [policy on] a large number of such sites, with a large productive capacity, should be completely different from the approach to [policy on] a small number of such sites.

'A certain quantitative leap took place: there were more of these sites, and they became much more powerful, but the attitude to how they were run deteriorated.'

'Why did this happen?'

'I think that a very strong element of inertia was responsible for this. The need for electrical power is great. It was necessary quickly to introduce and master new scales of power. And quickly: that means, not making any fundamental changes to projects which had already been agreed. The number of people busy with the preparation of installations and their running increased sharply. But the teaching and training methods could not keep up with the rate of development.

'It would be comparatively simple, if it was possible to imagine the enemy, let us say, in the shape of the nuclear reactor or in the shape of nuclear power engineering. But this isn't the case. And even if we reject this technical method and replace it with another, it still won't be ok. It will be worse. You see, the enemy isn't technology. It isn't a question of the type of airplane, of nuclear reactor, it isn't the variety of power engineering. If you look at this problem from a broad perspective, then the main enemy is the very means of creating and running power or technical processes, and the means is dependent on man. The human factor is the most important thing. If earlier we looked at safety technology as a means of protecting man from the possible influence on him of machines or some sort of harmful factors, then today we have another situation.

'Today technology must be protected from man. Yes, from man, in whose hands colossal forces are concentrated.

'Protected from man in any sense: from the construction engineer's mistakes, from the designer's mistakes, from the operator's mistakes, he who leads this process. And this is a completely new philosophy.

'Now what world tendencies can be traced? The number of accidents if we take the specific gravity per thousand men or another indicator is going down. But if it happens anyway with less probability, then its scale increases.'

'It's like an airplane: once fourteen people died in an airplane accident, now it's three hundred.'

'Quite correct. And here's the first conclusion: Chernobyl showed that humanity did not really rush with its change of approach to safety, to a philosophy of safety. We must bear in mind that not only the Soviet Union is lagging behind. This is a worldwide phenomenon. Consider the Bhopal, Chernobyl and Basel tragedies.

'It would be impossible, incorrect and foolish to deny the achievements of human genius. To deny the development of atomic power engineering or the chemical industry. But we must do two things: first, understand correctly the influence of such serious new machines and aspects of technology on the environment and, secondly, elaborate a system of interaction of man and machine. This is the problem not just of the man who works with such a machine, this is a far more general and important problem. After all, at times of such interaction serious catastrophes can occur, problems through oversights, stupidity, through incorrect actions. It isn't important who made the mistake: the head of the station or the operator.

'Today we must look for what is best in a system. The optimum in automation, the optimum in man's involvement in the processes, the optimum in solving all organizational and technical problems connected with such complex technological systems. At the same time we must create protective barriers, so far as this is possible, in case it is man who makes a mistake, and the machines become unreliable.

'Now I want, for the first time, probably, to express to you one, perhaps, unusual thought. So far we have discussed well-known things. Well, we all see, so to speak, with the naked eye, that there is at all stages of the creation of technology a certain incompleteness, even slovenliness in our work. At all stages—from the creation to the running. These are generally known facts, they are expounded in the decision of the Politburo of the Central Committee of the CPSU regarding the causes of the accident at the Chernobyl nuclear power station. And all the time I thought: why does this happen?

'And, do you know, I come to a paradoxical conclusion. I don't know if my colleagues will agree with me or throw stones at me, but I draw the conclusion that this happens because we have got too carried away with technology. We have become too pragmatic. With naked technology. This embraces many questions, not only of safety. Let us think for a moment: why, when we were far poorer, and the international situation was far more complex, why in a historically short period, during the thirties to the fifties, did we manage to astonish the whole world with the rate of creation of new types of technology and be admired for its quality? After all, the TU-104, when it appeared, that was a quality plane. The nuclear station which Igor Vasilevich Kurchatov and his companions created, that was both a pioneering and a fine decision.

'What happened, and why?

'The first test is to explain this by, let's say, subjective and organizational factors. But this isn't very serious. We are a powerful people and we have great potential. And every director, every organizational system at a certain historical stage applied various methods—some successful, others less so—but they couldn't have exerted such a great influence.

'And I came, roughly, to the following paradoxical conclusion: that technology of which our people is proud, which ended with Gagarin's flight, was created by people who stood on the shoulders of Tolstoi and Dostoevsky. . . .'

'That's a shattering conclusion in the mouth of a technological specialist.'

'But it seems to me that it is the correct conclusion. The creators of the technology of that time were educated in the spirit of the greatest humanitarian ideas. In the spirit of a beautiful literature. In the spirit of great art. In the spirit of a beautiful and correct moral sense. And in the spirit of a clear political idea of the structure of the new society, the idea that this society was the most advanced in the world. This high moral sense was there in everything: in the attitude of one person to another, in the attitude to man, to technology, to one's duties. All this was there in the education of these people. And technology for them was simply a means of expression of moral qualitites, placed in them.

'They expressed their moral attitude in technology. Their attitude to the creation and use of technology was the one which Pushkin, Tolstoi and Chekhov taught them to have toward everything in the world.

'And already in the generations that succeeded them, many of the engineers stood on the shoulders of the technocrats and saw only the technical side of things. But if someone is educated only in technical ideas, he can only reproduce technology and perfect it, but he cannot create anything qualitatively new, for which he can be responsible.

'It seems to me that the general key to everything which is happening is the fact that we have for a prolonged period been ignoring the role of the moral principle: the role of our history and of our culture; and this is one uninterrupted chain. All this has led, strictly speaking, to the fact that some of the people, in their positions, could have acted without due responsibility. And even one person, working badly, creates a weak link in the chain, and it breaks.

'In fact, if one listens to those directly guilty of the accident, then their objective was only the general good. To carry out their mission, to carry out their task.'

'Valerii Alekseevich, did they on the whole realize what they were doing?'

'They considered that they were doing everything well and correctly. And they were breaking the rules for the sake of doing it even better. That's how it seems to me.'

'But did they realize that they were breaking the rules concerning the running of the reactor?'

'They couldn't help but realize this. They had to. Because they were breaking fundamental, so to speak, orders. And some considered that this was safe, others that doing it like that was even better than in the instructions, because their objective, you see, was entirely honourable: to get themselves ready and without fail do what they had been charged to do, that very night, whatever the cost. Whatever the cost.

'It is true that this does not apply to those people who extremely irresponsibly permitted the tests and approved the programme to carry them out. The sense of the experiment was precisely in this. If the supply of steam to the turbo-unit stopped—that was an emergency situation—then the diesel generators at the stations had to be brought into play. They take up the necessary parameters to supply the block with electrical power not immediately, but after a few dozen seconds. In this case, the generation of electrical power has to be provided by the turbine, which has lost steam, but still turns by inertia. It was necessary to check if there was enough time for the spinning turbines to provide the needed power before the diesel generators started up. The programme to check this had been put together very carelessly, without any agreement either with the station physicists for the reactor builders, or the designer, or the representatives of the State Nuclear Power Inspectorate. But it had been approved by the Chief Engineer, and then was not controlled by him personally, but was changed and interfered with as it was carried out.

'The low technical level and the low level of responsibility of these people is not a cause, but an effect. The effect of their low moral level.

'Usually people understand as follows: an immoral person is one who allows himself to take bribes, for example. But that's an extreme case. Can you really call a person moral who does not wish to do his sketch better, does not wish to sit up at night, tormenting himself, looking for the most perfect solution? The person who says: "Why should I make a great effort when I can find a solution which seems satisfactory professionally, but isn't optimal, isn't the best?" And this is where the process of a spread of technical backwardness began. We will not cope with anything if we do not renew our moral attitude to the work that is being done, whatever sort of work it is: medical or chemical, biological or to do with reactors.'

'But how can this moral attitude be renewed?'

A sigh, and after a long pause:

'Well... I can't be a prophet.'

'And yet, Valerii Alekseevich. Imagine that you are Minister of Education or a person who decides the fate of schoolchildren. What would you do?'

'I've already answered this question in part: one must renew the sense of responsibility, of criticism, of a sense of the new. There was a time when certain external circumstances prevented this. But today we have a much more favourable period. Fortunately, nothing prevents us from renewing the best Soviet and national traditions in our multinational country. And no one prevents us either. But how can we do this? By increasing or decreasing the quantity of these or those things? I don't know, but I am sure that you should invite interesting people to the school. After all, our country has always been famous for the fact that the teacher is a person who from the moral point of view is most often his pupils' ideal.

'I also want to talk of the indivisibility of general and technical culture. These are indivisible things. If you have excluded some part connected with the history of our fatherland or our literature, if you have relaxed the attention you pay to something, this will without fail return like a boomerang because of the indivisibility of culture. Likewise, you can't devote everything to literature and art and forget technology. Because then we shall become an inept society. It is a natural problem—the problem of harmony.'

'Let us return to Chernobyl. How did you live this event as a person and as a specialist? Did you not have a guilt complex, not a personal one, but a physicist's guilt complex for what happened?'

'I would say that there was a feeling of anger. And displeasure that here, in our institute, where specialists expressed all the possible apprehensions and proposals, we turned out to be insufficiently strong and equipped to put the necessary point of view into practice. We wrote reports, we gave talks, we had a presentiment of the danger of a complication of technological systems if there was no change in the philosophy of their construction. There were recommendations prepared too. For example: the most important safety element would have been the creation of diagnostic systems. Among us were people who fought for these diagnostic systems, some of them were tested, we demanded that they be developed, everywhere we explained the danger of the fact that we did not have the computing power to build the necessary models and to evaluate the situation, or to train staff on simulators. But it turns out that they asked for little and explained badly. So in this sense there was a feeling of anger, or something similar. To get angry with physicists or all the more with physics, it's like beating a gutta percha copy of a boss with a stick, as they used to do in Japan. Physics is the leading edge of our science, it can't be guilty of anything. It is the people who exploit it who are guilty.

'How did I live through this as a person? On Saturday 26 April they took me off active duty. I simply flew there without prior preparation. None of us had expected an accident on such a scale. The station provided those of us in Moscow with the wrong information. We had contradictory information. According to one piece of information it was as if everything had gone wrong there: a nuclear accident, the threat of radiation, and a fire, every type of danger. Later we received notification that they were trying to cool things down, that is, they were making an attempt to control the reactor: that meant that the reactor existed and that at that time there were no particular problems. But when we arrived—that was on Saturday evening—I saw the purple glow. That, of course, astonished me and immediately showed me that the problem was serious. And later there was no time for emotions, you had to think on the spot, what, with what and how to make measurements, what measures to take, etc. That evening we only evaluated the radiation situation, for which the most active "dosimeter operator" was Professor Armen Artavazdovich Abagian. On the following day, when I went to the ruins of the reactor in an armoured troop carrier, then I had that sense of anger that I've talked about. And that feeling that we were unprepared for such a situation. There were no solutions and technical remedies worked out in advance. After all, what had happened? It had always been said that the probability of a nuclear accident was very insignificant. And the projects for the stations really took this insignificant probability into account. But then, it wasn't a zero probability. It followed from that that such an accident could happen once in a thousand years. But who said that this "once" could fall in our year, 1986? However, the possibility of emergency events before this unlikely event took place was not foreseen.

'And it is true that some time later, in Vienna at the IAEA

meeting, I was convinced that the whole world's science and technology, as the reality showed, was not very prepared either for such accidents...

'Moreover—perhaps this sounds paradoxical—hardly had the sharpness of the alert slackened, than I began to feel satisified at the work that had been done. In my opinion I am not alone, not at all alone in these feelings. Because conditions had been created in which real work was done—no bits of paper, no fuss, no conciliation. The Government Commission bore a very great responsibility. Particularly in those first days. It was later, when the situation returned somewhat to normal, that all sorts of conciliation appeared. And then it was as follows: everyone helped us, everything was under our control, but all the responsibility for the decisions taken lay on the shoulders of people who came there, and particularly on the shoulders of B.E. Shcherbyna. And this turned out to be very useful. The situation was dramatic, but in the circumstances of the independence which had been granted us, linked with the responsibility, a large number of people managed through their organized efforts both to limit the number of casualties and comparatively quickly to localize the scale of the accident.

'We had to solve scientific tasks there too. The first was the localization of the accident. We did not have algorithm for what happened in such situations. And the single field of action was in the sky, at a height of no less than 200 metres above the reactor. What could we do? The first thing we convinced ourselves of was that the reactor was not working. Neutron sensors in these gamma-fields did not work, all the neutron channels were incapable of action. So it was necessary to determine, through the correlation of short-life isotopes and their rate of emission, that there was no new production of rapidly decaying isotopes. We convinced ourselves that this was the case. The reactor was not working. But the graphite was burning and heat was being emitted. If the graphite was burning, that meant that down below there was some sucking through of air and that a certain cooling was taking place. And so the process could be stabilized in its natural state, nothing need be done and we could wait while the natural cooling of the reactor took place. True, we would have to wait a long time. What was good about this? The good thing was that the threat of a penetration of the lower soil layers of the Zone, the danger of a melting of the base, the pollution of the waters below the soil, all this would be eliminated automatically. And there would be no problems.

'But then, through the air basin with its aerosol combustion products, and with the raised temperature, the activity of the reactor would penetrate much further and the scale and intensity of the pollution would be more significant. Covering the remains of the reactor from above meant a reduced danger of pollution through the air, but it would worsen the escape of heat, and so create a threat of warming up again and of a slipping of the fuel mass back down. It was necessary to take a decision. And what we decided was as follows: we would cover the reactor with materials which would both filter and at the same time stabilize the temperature. Consequently, metals with low melting points (as they melt, the temperature does not rise), which protect from radiation, and carbonate, which recover the reactor's heat for their own decomposition and in this way release carbonic acid gas: all helps to stop the burning of the graphite.

'A problem without precedent in the history of the world had been solved.

'Traditional appliances, as a rule, were useless either because the places where the measurements had to be made were inaccessible, or because of the high temperatures and the radiation fields. Many specialists and organizations had to find, in the shortest possible time, both new methods and new technical means to measure and to keep the active elements in place, so that they were not carried away by the wind, for construction and deactivization. A great deal was done and, as we can now see, the objective was attained. Western experts will later call these methods innovatory and effective. One can simply regret that all this was not efficiently done before the accident, but after it. But during those first days one had to work intuitively.

'And the last thing I want to say is about the young people. Of course, we had to observe different situations, sometimes not very pleasant ones. But among the young people there were those who evoked nothing but enthusiasm. Among us a lot was written about the heroism of the firemen. Some people, when they read that, argued that they had, for instance, spent far too long, and in vain, in the conflagration and that they had exposed themselves to more radiation than they need have. But this was real heroism, and it was justified, because in the machine room there was hydrogen and oils... They did not permit the fire to spread; if it had, it could have led to the destruction of the neighbouring block. The first localizing step was done correctly.

'And what about the military airmen! That really was an exploit. They worked faultlessly both from the professional point of view and from every other one. There were also many young fellows in the chemical subdivisions. The reconnaissance work lay on their shoulders, and they carried it out fearlessly and precisely.

'You know, there was such harmony there. I cannot say that the young people worked more than the others, but it is a fact that they worked in an entirely worthy fashion. The physicists, both those from Moscow and those from Kiev, climbed into the same hell. I would say that the young people showed, in their work, high human and professional qualities.'

Vladimir Stepanovich Gubarev, writer, journalist, laureate of the State Prize of the USSR, author of the play *The Sarcophagus*:

'Everything that happened in Chernobyl and around it was for me very unpleasant. I consider that in the history of our country this is, in its significance, the third event.

'The first was the Tatar-Mongol yoke. We defended Europe from the hordes and from barbarism. The second was fascism. We saved Europe from fascism. And today we are securing the future of humanity at a very great price.

'The tragedy of Chernobyl, and in this is its peculiarity, is that we met the manifestation of nuclear energy precisely in the form of the so-called "peaceful atom." There will be no more such catastrophes. I can say that with absolute certainty. And the future of civilization is unthinkable without nuclear energy. But Chernobyl happened. So, when we build this future, we must bear in mind the lessons of Chernobyl. Before Chernobyl we moved toward this too lightly. And so, truly, we are paving the way to the future at a very high cost.

'I would be a very primitive person if I described in an artistic form the documentary events. It is clear that a great deal of what forms the basis of the play was born in Chernobyl, where I was a correspondent of the newspaper *Pravda*. But I can say with absolute certainty that I had no concrete person in mind. I strove to create images of types.'

From the play *The Sarcophagus* (*Znamia*, No. 9, 1986):

Sergeev: There we couldn't imagine for a long time what had happened; so, just in case, we didn't notify Moscow. We waited for something...

Bessmertny: I think it's a very serious accident. For some reason they're not saying anything on the radio.

Sergeev: An explosion though?

Ptitsyna: Of course. It's just that this was the last thing that some people wanted, and so they are trying to prove that the reactor collapsed without an explosion. A fire. Just a fire.

V. Gubarev:

'When I set to to write *The Sarcophagus*, it was a natural wish to make some philosophical sense of this event. I wanted to show that we live at a completely different time than we imagine. That we live in the atomic-space age, that it has its laws, its philosophy, its responsibility for the actions of men and their effects.'

From the play *The Sarcophagus*:

Bessmertny: But what swine, excuse the unliterary word, what swine switched off the emergency system?! I wanted to say that this is murder. Not suicide, but murder!...

The Physicist: ... The main thing for you is to clarify who took off the emergency protection.

Bessmertny: Who took it off? Who took it off? It was the system that switched off the emergency cooling. A system of irresponsibility.

The Operator: And we're all hurrying, all rushing, taking on obligations, you'd say, three months earlier than the time limit, two full days, and he asked four times for a counter, and no one hurried up there. And we're carrying out the bosses' requests... Why? They ask—silence, and we cheer!—and forward we go!.. And all for the sake of the report, the bonuses... Who needs this acceleration? It just the same as letting 100 kilometre an hour cars loose around the town, let them squash everyone, the main thing is to go as quickly as possible. They promised to let it go on full capacity immediately after the holidays. Two days earlier than the time limit. Taking on obligations in all directions... What are we, clowns?

The Physicist: So they took off the protection.

V. Gubarev:

'In *The Sarcophagus* there are three basic ideas. The first: if a man acts according to his convictions and views, if he avoids responsibility, then that man will live in a sarcophagus.

'The second idea: if people—both individuals and society as a whole—do not draw conclusions from the tragedy, then they end up in the sarcophagus.

'And the third idea: constantly in the play, like a refrain, words from the civil defence instructions are repeated, as a model of a nuclear war. I wanted to say: if humanity does not ponder on the lessons of the tragedy, it will be in the sarcophagus.

'This play was written in a week. It was in July, from 19 to 26 July. When I began to write it, I could no longer sleep, I could not talk, I slept for three hours out of twenty-four. There was no other way. You understand, now I value these people, wherever they live, whatever they do, whatever jobs they have, for their attitude to Chernobyl. If a man is indifferent, if this tragedy did not touch him, then such a man, in my opinion, is lost. For there are national tragedies, and this is a national tragedy, when every man must declare his attitude to that event. I want to look those people in the eyes who say that the play isn't necessary, that it's premature. Because if we do not pay attention to the terrible, if we do not give voice to a warning, then there will be no one to look at, there will be no one to read our plays and our literary works.'

From the play *The Sarcophagus*:

The Physicist: The main thing in this tragedy is its lessons. We have no right not to learn them... The history of humanity has not yet had such an experience. The explosion of the reactor and its effects. It is not excluded that this is an isolated case. More precisely—the first. It has to be the last. For this, study along all parameters. Scientific, technical, psychological.'

V. Gubarev:

'And most important is that these lessons were not in vain for our young people. After all, those who were born after 1961, after Iurii Gagarin's flight, naturally perceive that they were born in the atomic-space age. They are used to rockets being launched. But they must understand one thing: if they live in this age, the level of their knowledge and culture must be far higher than their parents'. For they assume the control of a fundamentally new technology. And tomorrow they will create it themselves. And they sometimes perceive all this as their due, as somehow given. Like a car in the street. Or like a television. But this is the most complex technology. And very dangerous. It demands from man a new level of thought and of knowledge, and, most importantly, a new attitude toward it.'

Robert Gale:

'Chernobyl has provided us with many lessons to learn. One of them is the need to learn to co-exist with nuclear energy. There is no alternative. We live in the nuclear age and must make peace with it. In the USA we receive almost 17 per cent of our electrical power from nuclear power stations. In some Western European countries this figure has reached 60 and 65 per cent. By 1990 on the Earth there will be nearly 500 nuclear reactors. In other words, there can be no question as to whether we should enter the nuclear age or not. We are already in it. And so, in the use of nuclear energy, a high degree of responsibility, precision and caution is necessary. If we analyze the causes of all the accidents that have occurred in the USA and the USSR, we can see that they had their origin not in nuclear energy itself, but in mistakes made by man.

'One more lesson is that accidents like that at Chernobyl touch not only that country in which they occur, but several neighbouring countries. So in the case of such accidents help must be given not only on a national level, but also on an international level. We must understand that we depend upon each other and that nuclear energy and nuclear arms expand the geographical frontiers.

'And, finally, the last, most important lesson. If we compare it with the conscious application of nuclear weapons, Chernobyl can be termed a minor incident. But if a comparatively little accident has cost the lives of many people, the extensive joint efforts of doctors and two billion roubles, then what can be said of the wartime application of nuclear weapons? We, the doctors, will then be powerless to help people.

'This must never be forgotten.

'Chernobyl is humanity's last warning.'

One cold November morning, when the wet snow was falling on the clayey earth, I arrived at the suburban Moscow cemetery of Mitino. Not far from the entrance, to the left of the main avenue, stretched neat rows of similar graves. White marble headstones, golden inscriptions. Different dates of birth, almost all the dates of death in May 1986.

The heroes of Chernobyl. The victims of Chernobyl. It is possible that among them are also those guilty of Chernobyl. Death made them equal and gave us, the living, the right only to one feeling: a feeling of deep sorrow at the loss of these young human lives.

I bowed to their ashes (in the meantime I had, it is true, to show my writer's card to the duty policeman, as if there was something suspicious in what I was doing) and I drove off with heavy thoughts about the time lived by us after Chernobyl. This accident with its pitiless X-ray radiation had in a second lit up our national, state mechanism. More clearly than ever there appeared, on the severe screen of Chernobyl, both our great internal forces and reserves (after all, we can, when we want, solve any problem!) and our serious, long-established ailments, which can in no way be packed away in the placid formula of the past year: 'isolated atypical shortcomings.'

Dr. Gale is right! Chernobyl has struck us as the last warning: to humanity, to the country, to every one of us—young and old, chief or subordinate, scientist or worker.

All of us.

The last warning.

I do not want to comment any more on anyone, I do not want to prove, to explain, to persuade, to shout and to observe, because they beseech us and warn us, these people so different and unknown to each other: the Russians, Ukrainians, Belorussians, Georgians, Poles, Americans, and the golden-haired, gentle Aneliia Perkovska, who, having dispatched the children of Prypiat to the pioneer camps on 11 May, lost consciousness and fell and was taken to hospital in a serious condition; and Leonid Petrovich Teliatnikov, with whom I had the chance to speak in one of the Kiev hospitals—at that time he was feeling better already, his head was covered with fine, dark-red hair but all the same he acknowledged that he slept badly at nights and was pursued by visions of the fire; and the 'Man of the United States of 1986,' the brilliant Doctor Robert Gale, who touched our life and our misfortune; and the future heart surgeon Maksym Drach, who grew up by so many years in May 1986; and Academician Valerii Alekseevich Legasov, who spoke such bitter and pitiless words about the moral causes of all our misfortunes.

They have said everything, and their words do not require extensive commentary.

And if their voices and their truth are not heard, if everything remains as it was, if we learn 'anything and anyhow,' if we work as we used to work—anyhow, very slowly—if those who get on in life are the blindly loyal, cynical and illiterate 'yes-men' and not sensible and decent people with their own views and convictions, and if only total subordination to orders and not the creative comparison of different, freely expressed opinions are the highest virtue on the various hierarchical steps of the state, then all this will mean that we have learned nothing and that the lessons of Chernobyl have passed in vain.

And then there will be new Chernobyls, new Admiral Nakhimovs [a tragedy that occurred on the Black Sea in late 1986], new bitter shocks in our life.

The warning of Chernobyl. It happened that I was watching the TV film *The Warning*, shown in February 1987 on central television, in one of the Kiev hospitals together with those who had been working in the Zone, and now were under observation. All the department came together for the television, and although they were all different, people who did not know each other, that evening they were all united by the TV screen and by heavy memories of what they had lived through. I recollected my childhood, how, in an unheated cinema auditorium in 1942 in Saratov, the hungry, weary people had watched the documentary film *The Rout of the German Fascist Forces near Moscow*. They watched with pain and hope, sorrow and faith.

The times have changed, the historical circumstances have changed, the people have changed, only the expressions on the faces have remained unchanged: the same pain and hope. Beside me there sat young fellows in hospital pyjamas, the Ukrainian TV operators Iurii Koliada, Serhii Losiev, Mykhailo Lebediev, the director Ihor Kobrin, the commentator Hennadii Dusheiko. They looked intently at the frames of the chronicle of the events at Chernobyl. Already a bit like a 'Who's Who?,' but they knew at what cost those frames had been obtained. Iurii Koliada was the first television operator in the world who in April 1986 managed to film the ruins of the reactor. Every step closer to the site in those days cost dozens of rays. The people around me knew the cost of Chernobyl: over fifty workers of Ukrainian teleradio alone—television camera operators, radio journalists, commentators, sound engineers, drivers—were forced to go through a medical examination and some had to go for treatment in sanatoria. One of the leading and most fearless operators of Ukrainian television, 49-year-old Valentyn Iurchenko, suddenly died in the autumn of 1986. And although the cause of death (a heart attack) was not directly connected with the Chernobyl radioactivity, who can deny the role of stress and nervous hypertension which this courageous man went through in the hot summer days of 1986? That is the price which paid for the truth about Chernobyl, a truth which in itself had become the most serious warning to all of us.

Chernobyl began a special countdown for humanity.

. . . In empty Prypiat we went to the town's central defence point. The duty militia officer was sitting at a signals desk. In the adjoining room the patrol leader was rebuking the sergeant for something. It was all so normal. On a plywood panel in front of the duty officer hung bunches of keys. The name of the street, and a yellow bunch of entrance keys for the buildings. By their quantity you could see on which street there were more houses, and on which less.

Well, I would not like, at the central point of the Martian guard of the Earth (militia or police, it doesn't matter), there to be bunches of keys to empty and forever abandoned countries. I don't want, somewhere in a common bunch with the name 'Europe,' there to glitter the little key to my land, Ukraine.

As a symbol of that terrible world, in which we were last year, there hangs in my garage a white suit, given to me in Chernobyl. According to the rules, probably, I should have thrown it away, after all, I wore it in the Zone; but I can't. It is dear to me like a memory and ominous like a warning. And when, in the evening, I turn on the headlamps and drive into the garage, there appears before me a blindingly white spectre, a spectre which today wanders Chernobyl's fields and Kiev's apartments. . .

Enough of that!

So I want to end my story with one idyllic memory: after everything that I saw in the Zone and around it, after the dead silence of the abandoned villages (I don't know why, but the village cemeteries, those 'shadows of forgotten ancestors,' where the living will no longer ever return, moved me more than anything), after the hospital wards and the expressions of those who lay on a drip, after the leaps of the needles on the dosimeters, after the danger which lurked in the grass, in the water, in the trees, at the end of May I drove out of Kiev for two days. I drove at high speed eastward along the empty Kiev-Kharkiv highway, stopping only at the barriers, for the dosimeter check.

I was driving to Myrhorod, to see my daughter and granddaughter. That same Myrhorod of which Nikolai Vasilievich Gogol wrote:

'That wonderful town of Myrhorod! What buildings there are here! Straw roofs, reed roofs, and even wooden roofs; a street to the right, a street to the left, everywhere a fine fence; hops twine along it, pots hang on it, behind it the sunflower shows its sun-like head, the poppy shows red, the fat melons peep out... Luxury!'

How long ago that was! From what a naive and serene past those words came. But even in May 1986 Myrhorod was wonderful. Wonderful in that there was no radiation—well, perhaps a little higher than usual. And no one advised people to close their windows.

The early May evening approached, when the air in Myrhorod is full of the lazy fragrances of the earth which has grown languid over the day. I went down to the bank of the little river Khorol, lay down in the grass, and half-closed my eyes. I heard nearby the croaking of mating frogs, I felt the freshness of the grass and the closeness of the water. On the opposite bank cows mooed, waiting to give their hot milk to the tin buckets. And suddenly I understood what happiness was.

It is grass in which one can lie down without fear of radiation. It is a warm river in which it is possible to bathe. It is cows whose milk can be drunk without worry. And it is the little provincial town, living its measured life. And the sanatorium, along whose avenues those who are resting slowly walk, buying tickets for the summer cinema and making friends—this is happiness. But not everyone understands this.

I felt myself to be an astronaut who had returned to Earth from a distant and dangerous journey into the anti-world.

At this moment an acquaintance of mine called me and handed me a plant, torn up by its roots. Nothing remarkable—coarse dark-green leaves and a thick stem, and as it were painted a little with violet ink. This plant was called *chernobyl.* It had a bitter taste.